QUARKS

The Stuff of Matter

HARALD FRITZSCH

ALLEN LANE

ALLEN LANE
Penguin Books Ltd
536 King's Road
London SW10 0UH

First published in German as *Quarks: Urstoff unserer Welt*
Copyright © by R. Piper, Verlag, Munich, 1981

This Translation first published in Great Britain, 1983
Translation copyright © Basic Books, Inc., 1983

ISBN 0 7139 15331

Translated by Michael Roloff and the author

To Brigitte

Contents

Contents

Foreword

Humankind has been obsessed with a concern for a world view since we have had records, and presumably long before then. The earliest puzzles concerned the cycle of day and night, the cycle of the seasons, and practical science that would diminish irrational human fears and insure the bounty of nature. But less applied science was also prominent; and the dreamers among our remote ancestors studied the night sky and the infinite variety of matter in its natural environment, and invented models in their quest for an underlying coherence.

The belief that there is, in fact, a key to the functioning of the physical world, and that this key will yield to human thought, was given credence with the success of science beginning in the sixteenth century. The combination of natural philosophy and practical arts gradually accelerated progress, as science led to invention, and invention, in turn, gave increasingly more powerful tools to science.

Professor Fritzsch's book on the state of particle physics summarizes current thought on the subject; this current thought is the culmination of a series of episodic events in which violent intellectual revolution alternated with steady, undramatic advances along the entire front between comprehension and ignorance. This book is an attempt to describe, in layman's terms, the events of the past several decades. By 1940, physics had become comfortable with the revolutions taking place between 1920 and 1930 in relativity and quantum mechanics. The next development was the postwar application of new technology (with the blessings of impressed governments) to the construction of powerful instruments, particle accelerators. These are properly credited here with opening a rich domain of the microscopic universe to observation.

It is appropriate in this survey to concentrate on the evolution of ideas and experimental results. Underlying both theoretical and experimental happenings are a series of brilliant inventions in both the accelerator domain and the domain of particle detectors. Notable developments include the invention of the new accelerator principle, strong focusing, by Ernest D. Courant and Hartland S. Snyder in the early 1950s; this generated a series of accelerators, beginning with Robert R. Wilson's Cornell electron synchrotron and the Brookhaven AGS and culminating in the CERN SPS and Fermilab, whose photographs appear in this book. The pioneering ideas of Gerard O'Neill, Laurence Jones, and Burton Richter gave us colliding beams, examples of which are the PETRA, PEP, and ISR colliders. The bold application of superconductivity by Robert Wilson at Fermilab gives us a window on the 1000 GeV region. In the evolution of particle detectors, the notable events were the invention of the bubble chamber by Donald Glaser and the development by Georges Charpak of a series of devices for electronic registering of the precise location of particle trajectories. Particle physicists invented the digital circuitry that formed the basis of modern computers, and, in turn, profited enormously from the explosive development of commercial scientific computers. It is a source of endless wonder that the fashioning of such complex technological objects is a necessary ingredient in our appreciation of so abstract a concept as the confinement of quarks or the existence of atoms of gluons.

The postwar accelerators provided enough energy to probe the nucleus deeply. The first results were astonishing; new particles were discovered, wrenched out of the strong force field of the nucleus. These proliferated to such an extent that, in naming them, we were in dire danger of running out of Greek letters. As Professor Fritzsch describes with detail and clarity, recognition of the organization of the properties of these objects provided the clue to their underlying structure. These properties were subtle and difficult to measure, and required the deployment of detection instruments and strategies of great ingenuity, a major story in itself.

The quark substructure, the leading characters in this story, first took shape as a logical speculation whose reality was questioned, even by those who proposed the quark hypothesis. The crucial experiments on "deeply inclastic scattering," in which the small, hard objects within the proton were first detected by probes of electrons and neutrinos, indicated the reality of quarks. Important confirmation came from the muon pair experiments, quite a different process but providing the same numerical information on the quark constituents of colliding protons. But the final clinching element that established the quark theory was, in fact, the discovery of new quarks—the charmed and bottom quarks—in the mid-1970s. There is a marvelous interplay between theory and experiment unfolding in this story, so that experiment drove theory and theory suggested and interpreted experiment.

It is useful to note that similar actions were taking place in astronomy, where new forms of observation led to dramatic conclusions about the form, structure, history, and mechanisms of the physical universe. In his last chapter, Professor Fritzsch gives us a tantalizing hint as to how the two subjects, the microworld and the cosmology of the origin of the universe, have merged in their use of the grand unified theories.

The story told here is clearly still unfolding. The synthesis described in this book still has vast possibilities for surprise. As Professor Fritzsch points out, we have come to a point (as of this writing) where there is urgent need—in the seesaw history of the interplay of theory and experiment—for new data at higher energies. The goal is clear and tantalizing: a complete and consistent theory that fully explains how the universe works. Nevertheless, looking back, we can reaffirm the rhetoric of Jacob Bronowski:

Physics in the twentieth century is an immortal work.

Foreword

The human imagination working communally has produced no monuments to equal it, not the pyramids, not the *Iliad*, not the ballads, not the cathedrals. The men who made these conceptions one after another are the pioneering heroes of our age. (*The Ascent of Man,* 1973)

Once perfected, the new world view will enable scientists to understand the workings and history of our cosmos back to the fireball that started it all, to account for all observations in the present, and to predict how the universe will evolve in the future. This book undertakes the important task of rendering this process comprehensible to the general public who, after all, has paid the bill for this research.

LEON M. LEDERMAN
Director, Fermilab
Batavia, Illinois
October 1982

QUARKS

Introduction

Physicists are accomplishing extraordinary things in this century. Using both theoretical and experimental findings, they have been able to decipher the details of the universe—from the most immense distances to the smallest ones. Modern astronomers, gazing at stars and galaxies far away from the earth, can reconstruct what occurred only minutes after the creation of the universe, billions of years ago. Too, they have discovered new, previously unimagined objects—pulsars, quasars, and black holes. At the other end of the spectrum elementary-particle physicists are investigating the structure of objects so small that the size of a mere atom begins to seem huge. New and strange objects have also been found at this infinitesimal level—the strangest of all being the quarks, which are the subject of this book.

Though the extraordinary discoveries in physics made in recent times have required immense public funding, physicists have not been too successful in communicating their specialized knowledge to the interested public that provides the funding. One aim of this book is to show that anyone with an elementary knowledge of physics can comprehend what physicists have accomplished in the last thirty years. Following the physicists' methods of theorizing, experimenting, and sometimes ingeniously guessing, readers should find that modern particle physics is by no means as daunting as they may have supposed. What can be daunting, admittedly, is the sheer newness or unfamiliarity of the concepts and processes. But once having grown accustomed to them, the reader will realize that physics, if not precisely commonsensical in all respects, is fundamentally simple, orderly, and comprehensible. I hope that readers who follow the trail of these discoveries will share the excitement and wonder that physicists feel when they have made them.

About 500 years ago, a new dimension was added to man's age-old attempt to satisfy his insatiable curiosity about the nature of the universe—people began to verify their scientific hypotheses by stringent tests. It was Galileo Galilei who ushered in the era of modern science 400 years ago when he conducted the first verifiable experiments during his investigation of freely falling bodies. Compared to Galileo's throwing rocks off the leaning tower of Pisa, our using gigantic particle detectors to penetrate to the heart of matter seems to represent something of an entirely different order. Yet the principle behind both kinds of experiments, and the excitement of discovery they engender, remains the same: when we conduct an experiment we put a question to nature, and nature is forced to yield an answer. Frequently the answer is rather complicated and we have enormous difficulty understanding it. Sometimes, however, the answer is simpler than we anticipated. When insights suddenly appear and new ideas are born—these are the great moments of science, when it becomes as sublime as a great work of art.

At the turn of this century two new theories, quantum mechanics and relativity, had a revolutionary effect on the development of physics. Einstein's theory of relativity, first formulated in 1905 while he was an employee of the Federal Patent Office in Bern, changed our whole approach to space and time. The classical notions of space and time, which were first formulated mathematically by Isaac Newton and which we still use every day, become inapplicable when we are forced to think in terms of speeds such as the speed of light, which is 300 000 km per s. One of the consequences of the theory of relativity is that nothing can move faster than the speed of light; another is the nonconservation of mass. Normally we assume that the mass of an object is equal to the sum of the masses of its parts. If we take two steel balls, each having a mass of 1 kg and put them together, we expect the total system to have a mass of 2 kgs. This is not the case, however, when relativity becomes an important part of our consideration. Relativity theory tells us that mass can be destroyed and created. For example, we can take a neutron and a proton (particles that will be discussed in detail) and form what is called a deuteron. It turns out that the mass of a deuteron is slightly less than the sum of the masses of the proton and the neutron.

Whereas the theory of relativity has changed our concepts of space and time, quantum theory has revolutionized our thinking about natural events. It is not possible, within the framework of quantum theory (developed by Max Born, Werner Heisenberg, Pascual Jordan, and others in the 1920s), to make statements with absolute certainty; we can only make statements about probability. For example, we know that within a certain time period neutrons decay into protons and some other particles. Yet it is impossible to predict precisely at what time a given neutron will decay. Specifically, this means that of 1,000 neutrons, for example, half will have decayed within eleven minutes, but we cannot say which half. If we wait a further eleven minutes, only half of 500—250 neutrons—will still be alive, and so forth. The laws of quantum theory, therefore, permit us to make statements about quite a few things (as about neutrons just now), but they do not permit us the luxury of making definitive statements about any specific process. The probability of the decay of a single neutron does not increase with time; a neutron does not "age." The probability of its decay within any eleven-minute period remains the same, that is, 50 percent. Something like roulette prevails here: everyone can calculate the chances of winning, but no one is assured of coming out a winner.

Critics have repeatedly tried to interpret the uncertainty of quantum theory as a consequence of our ignorance of elementary processes. It is possible to think of the neutron as a rather complicated physical system that will break down when a certain set of events occurs. Ignorance of what leads to the decay might lead an outside observer to venture guesses in terms of probabilities. However, some lucky physicist with the right microscope might be able to reach beyond probability and discuss neutron decay with absolute certainty.

Today, however, we feel that our ignorance is not due to our inability to observe what really occurs during elementary quantum processes, but that the laws of quantum theory set an absolute limit on our ability to predict. We shall *never* be in a position to predict when a specific neutron decays. In this sense, the limits of quantum mechanics are just as strict as those of the theory of relativity, which set the speed of light as the ultimate velocity. Of course, this inability to predict with greater certainty than that permitted by quantum theory has troubled many physicists, including some of the theory's founders. Albert Einstein expressed his doubts with the famous dictum, "I shall never believe that God plays dice with the world."

Nonetheless, in the meantime, essentially all physicists today accept quantum theory, if only because of its great success in explaining the dynamics of atoms, atomic nuclei, and elementary particles. Our need to resort to probability to interpret quantum mechanics demonstrates the fundamental limitations of concepts based strictly on our experience of the macroscopic world. Realizing that an average-sized rock is generally at least 10^9 times larger than a single atom helps to explain why we have some difficulty translating concepts born of the macroscopic experience into the terms of elementary particles. The probability interpretation of quantum mechanics is the compromise we have to accept if we want to use our macroscopic concepts of space, time, and velocity when referring to something as tiny as an electron.

For example, quantum theory tells us that we cannot simultaneously determine the location and velocity of one specific elementary particle, for example, a proton. In the macroscopic world we encounter no problems of this kind. Everyone knows what we mean when we say that a car crosses Times Square in New York at 30 km/h: the car has that velocity and is at that location at the time of observation. We can measure the velocity of the car by using radar (whose waves are similar to light waves but longer). The radar waves hit the car and reflect back at us, and an analysis of the reflected radar waves provides us with the desired information about the car's velocity. The location of the car also can be fixed using ordinary light. Strictly speaking, our eyes receive light signals that our brain then analyzes to obtain information about the location of the car. Such signals, either radar or light, constitute a special form of energy and momentum.

Radar, light, and other electromagnetic waves transfer momentum and energy through space. For example, the radar signal that hits the car from behind will transmit energy and momentum to it, causing it to accelerate. As a result, the velocity of the car will be increased. However, the momentum that the wave contributes to the car's acceleration is so minute that we may disregard it. Because we may neglect such minute contributions in the macroscopic world, we can determine both the exact velocity and the exact location of the car at a specific time. Beginning in the sixteenth century, physicists developed the entire science of classical mechanics based on this possibility.

The principles of classical mechanics, however, do not apply to elementary particles. Determining the location and velocity of an atom or an elementary particle presents a far greater problem. For example, let us consider a proton located at a certain point in space. In an experiment to determine the location of the proton, we shine light on it. The proton will reflect the light waves; and analyzing the reflections of these waves, we can draw certain conclusions about the location of the particle. Unlike the car, however, the proton will change its velocity drastically when hit by the light. If the proton was at rest when the light wave struck, it will now be moving in a certain direction because the light wave has transferred momentum and energy to it. For this reason it is impossible to determine simultaneously both a proton's speed and its location with complete accuracy.

In a certain sense, the notions of velocity and location are complementary. The location of an elementary particle can be determined with great precision, but the very fixing of its location throws its velocity into doubt. The reverse happens when we fix a particle's velocity with precision. Once having done so, we can give only a poor account of its location.

Werner Heisenberg, in the 1920s, was the first to recognize the importance of the complementary relationship between various physical quantities such as velocity and location. He discovered that there exist well-defined relationships between the various limits of uncertainty. These uncertainties of physical quantities are governed by a constant determined by Max Planck at the beginning of this century. This quantity, called Planck's constant and designated by h, is of fundamental importance to physics. It's value is $h = 6.6 \times 10^{-34}$ W s^2. Readers unfamiliar with physical quantities will have little use for this numerical value. I mention it only to indicate how minute the constant is, which explains why we can neglect quantum mechanical uncertainties in our macroscopic world. We simply do not notice them. For example, there also exists a quantum mechanical uncertainty as to the location and speed of a car, but this uncertainty is extremely small, much smaller than the radius of one atom, and therefore we may neglect it. Classical mechanics still rules the macroscopic world in which Planck's constant might as well be zero.

Planck's constant, however, is extremely important to our understanding of atomic structure. Combined with the theory of relativity, quantum theory has enabled us to understand the structure of atoms. Today we can state unequivocally that the physics of the atom is understood. A few details need to be cleared up, but that is all. That is why basic research in physics since 1950 has turned toward an investigation of the constituent parts of atoms, of elementary particles like the electron and the proton. Rather we should say so-called elementary particles, since the proton, as we shall see, is considerably less than elementary. Important and fascinating discoveries have been made in high-energy physics, especially since 1969, and today it appears that physicists are about to take the important leap toward a complete understanding of matter. This book describes the developments that have substantially changed physics during the past ten years. Possibly the developments of the 1970s will be as important to the future of physics as the development of the quantum theory was in the 1920s. In any event, most physicists feel that something important occurred during the last decade, something worth transmitting to a larger public.

This book begins with a brief summary of the state

of atomic physics at the turn of the century and then describes the origins of particle physics. Its chief aim, however, is to introduce the reader to the concept of quarks as the fundamental constituents of matter. Though this is not a book on the history of the development of particle physics, it will often follow the historical route in introducing new concepts and ideas.

Since the general topic of this book is high-energy and subnuclear physics, this introduction would be incomplete if it failed to mention the main laboratories where the experiments are being conducted. High-energy physicists explore physical phenomena that occur at extraordinarily high energies—energies much larger than those we have become familiar with through atomic and nuclear physics. In a certain sense, high-energy physics tests our theories under very extreme circumstances, to the point where the theories break down so that heretofore unknown laws of nature emerge. In high-energy laboratories, particles such as protons and electrons are accelerated to enormous energies so that they move essentially at the speed of light. For example, the modern electron accelerators can accelerate electrons to a velocity of 0.999 999 999 86 of the speed of light. Subsequently the particles collide with other particles and sophisticated detectors observe what happens in these complicated reactions.

Now, research in high-energy physics has become so expensive that small countries can no longer afford to maintain their own research facilities. For this reason, work in experimental particle physics is now frequently conducted at large, sometimes international, centers. The leading laboratories where protons are accelerated to high energies are the European Nucle-

Figure I.1
The European Nuclear Research Center (CERN) near Geneva.
The CERN site is in the foreground. The dotted circle marks the
position of the 400-GeV Super Proton Synchrotron, 2.2 km in
diameter, built in a tunnel below ground. In the background, Geneva Airport, the city, the end of the lake, and on the horizon,
Mont Blanc.

Introduction

Figure I.2
The Fermi National Accelerator Laboratory (Fermilab), Batavia, Illinois. The largest circle is the main accelerator. Three experimental lines extend at a tangent from the accelerator. The sixteen-story, twin-towered central laboratory is seen at the base of the experimental lines.

ar Research Center (CERN) near Geneva (figure I.1) and the Fermi National Accelerator Laboratory (FNAL) near Chicago (figure I.2). The most important discoveries of the 1970s, however, were made at laboratories that accelerated both electrons and positrons (the antiparticles of electrons) to high energies. The leading laboratories in this field are the Stanford Linear Accelerator Center (SLAC) in Palo Alto (figure I.3) and the German high-energy physics laboratory, Deutsches Electronen-Synchrotron (DESY) in Hamburg (figure I.4). I have visited these institutions

Figure I.3
The Stanford Linear Accelerator Center near Palo Alto, California. The long, linear tunnel that houses the vacuum pipe in which the electrons are accelerated crosses the freeway from San Francisco to San José.

Introduction

Figure I.4
The German High-Energy Physics Laboratory (DESY) in Hamburg.

many times and have profited greatly from their research atmospheres. Had it not been for stimulating discussions with colleagues at CERN, FNAL, SLAC, and DESY, it would never have entered my mind to write this book. In particular, I would like to thank my friend and collaborator Murray Gell-Mann for the numerous conversations we had on the goals and wonders of particle physics.

I

A Look Inside
the Atom

Today we know that we and everything around us consist of atoms or of compounds of atoms. In general, atoms do not appear as isolated entities but are bound together into molecules, and it is in this molecular form that the different kinds of matter manifest themselves. Physicists today are interested not in single atoms or molecules, but in the substructure of the atom and in the various particles of which it consists.

The concept of the atom, which dates back to the ancient Greeks, was applied by chemists in the nineteenth century to try to fathom the laws underlying various chemical interactions. To these chemists, atoms were small bits of matter of a certain mass and shaped like billiard balls, each with a radius of about 10^{-8} cm (so small, that is, that it would take 100,000,000 atoms to make a line 1 cm long).

A Look Inside the Atom

The question asked by early twentieth-century physicists was, What is the structure of the atom: is the matter inside atoms distributed more or less uniformly or do atoms have a specific structure? To find the answer to this question, early physicists conducted experiments that resembled those presently being undertaken in large accelerator laboratories around the world. Of course, particle accelerators did not exist at the time; all early scientists could do was study the trajectories of alpha particles, (hereafter referred to as α particles, which are positively charged subatomic particles emitted by some radioactive substances) when these particles penetrated certain materials, such as, for example, a thin piece of metal foil. If matter were more or less uniformly distributed inside the atom, α particles would penetrate the foil without being deflected, much like a bullet passing through water. The experiments provided some surprising results, however. Although a bullet traveling through water will slow down after a while because of the resistance of the water, it will not change its course abruptly. With α particles however, there is a change in direction. As expected, the α particles frequently went straight through the metal foil, losing part of their energy in the process, but sometimes, however, they changed direction quite unexpectedly (figure 1.1). Exact measurements of the changes of direction showed that the α particles seemed to be colliding with tiny objects inside the atoms—objects far smaller than the atoms themselves, that is, with a width of approximately 10^{-12} cm, which amounts to $1/10\,000$ of the diameter of the atom itself. If we think of the matter at the center of the atom as being the size of an apple, then the entire atom has a diameter of

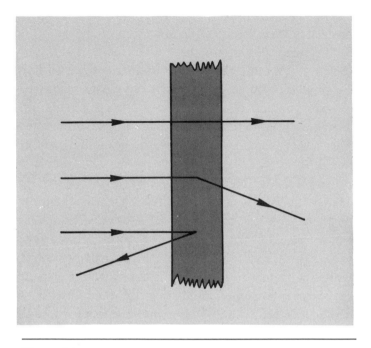

Figure 1.1
A beam of α particles passing through a thin metal foil. Some-times the α particles scatter at a sharp angle.

about 1 km. It was also discovered that the center of the atom, now called the atomic nucleus, contains most of the atomic mass and only a tiny fraction (less than 0.1 percent) of the total atomic mass is distrib-uted outside the nucleus.

The results of these scattering experiments finally led Ernest Rutherford and Niels Bohr to propose an atomic model in which the nucleus carries a positive electric charge and is surrounded by a cloud of nega-tively charged electrons. Electrons are exceedingly light elementary particles whose mass, expressed in energy units on the basis of the equivalence of mass and energy (from Einstein's relation $E = mc^2$) is about 0.5 MeV.*

The electric charge of an electron is an important physical constant. Since all electrons have precisely the same electric charge, we say that electron charges are quantized. The fact that all atoms are electrically neutral implies that the electric charge inside the nucleus exactly cancels the electric charge in the electronic cloud that surrounds it.

Careful investigation of the structure of the atomic nucleus during the first half of this century revealed that the nucleus consists of positively charged protons and electrically neutral neutrons (both called nucleons). The electric charge of the proton is exactly equal in magnitude to the charge of the electron, but is of opposite sign. The protons and neutrons are heavier than electrons. Their mass corresponds to approximately 940 MeV, which makes them about 2000 times heavier than electrons. The mass of the neutron is 1.3 MeV greater than that of the proton.

It is indeed remarkable that the electric charge of the proton and that of the electron are exactly equal in magnitude because in nearly every other respect electrons and protons are entirely dissimilar. In principle, the world could exist if the electric charge of the proton and that of the electron were unequal. In such a world, however, atoms would have a net elec-

*The abbreviation *MeV* stands for *megaelectronvolt,* and 1 MeV is equal to 1,000,000 electronvolts—the energy acquired by an electron passing through a voltage difference of 1 V. Later on, we shall often use the unit GeV—gigaelectronvolt—which equals 1000 MeV. It is of course also possible to express energy in units of mass. For example, 1 MeV corresponds to a mass of 1.8 × 10^{-28} g. It is apparent that the electron, with a mass of 0.9 × 10^{-27} g, is a very light particle indeed.

tric charge and, as a result, the structure of matter would be radically different. The proton and electron must have something in common that makes their charges equal in magnitude but opposite in sign despite the fact that the physical properties of the two (mass, behavior, and so forth) are so very different.

Physicists have, of course, been searching for the common denominator between electrons and protons, so far without results. It appears that electrons and protons have nothing in common that would allow us to deduce the equality of their electric charge. Even today we are puzzled about this. We shall return to this fundamental problem of modern physics later on.

By far the simplest atom is the hydrogen atom. Its nucleus consists of just one proton, its atomic "cloud" of just one electron. The electron moves around the proton in an essentially circular orbit. This at least was the picture Rutherford had in mind when he carried out his famous experiments with α particles. Rutherford's model of the atom, however, contained a fundamental problem. According to the laws of electrodynamics, an electrically charged object moving in a circle is supposed to emit electromagnetic radiation. Such radiation—light, for example—is a form of energy. A moving electron constantly emits energy, which means that the electron should come closer and closer to the proton and finally plummet into the nucleus. This, however, is roundly contradicted by what we observe. We can quite easily calculate how long a hydrogen atom would exist before its electron fell into the nucleus: the required time is short indeed—not even 1 s. Yet we are well aware that hydrogen atoms exist for much longer than this.

Although recent speculation on the dynamics of el-

A Look Inside the Atom

ementary particles suggests that the proton might not be infinitely stable (a phenomenon we shall discuss in detail later on), for our purposes here hydrogen atoms exist for an infinitely long time, and we therefore conclude that there must be some mechanism that keeps the hydrogen electron in its orbit so that it does not lose energy. This mechanism, it turns out, is provided by the theory of quantum mechanics, which states that electrons are permitted to move only in specific orbits (called stationary orbits). An electron in such an orbit emits no radiation, and the orbit is dynamically stable. Furthermore, quantum mechanics implies that there are a large number of possible stationary orbits (figure 1.2), each of which corresponds to a

Figure 1.2
A representation of the hydrogen atom with the proton, constituting the atomic nucleus, at the center. Normally the electron moves in the orbit having the lowest possible energy (solid line). If we supply the electron with energy, for instance, with electromagnetic radiation, it will move to an orbit with greater energy (dashed line).

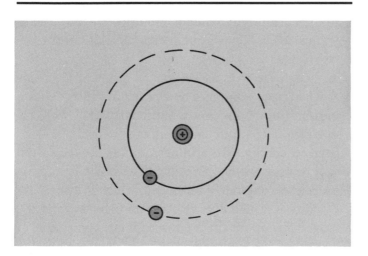

well-defined quantized energy. Specifically, there is a stationary orbit of lowest energy, called the ground state of the hydrogen atom. An electron moving in the ground-state orbit will stay there forever, if it is not disturbed, since there is no orbit with a lower level of energy. However, if energy is transmitted to an electron moving in the ground-state orbit (by electrodynamic radiation, for example), the electron can leap into an orbit corresponding to a higher level of energy. In such a case the atom is called an excited atom; and to excite an atom means to transmit to it an amount of energy that is exactly equal to the energy difference between the excited orbit and the ground-state orbit.

An electron moving in an excited orbit stays there only for a very short time, about 10^{-8} s, and then falls back to the ground state. It is at this moment that the electron emits energy in the form of electromagnetic radiation. The amount of this energy is the difference in energy between the two orbits. This energy might be in the form of light—for example, the light generated by a neon sign, which is produced when the electrons inside the neon atoms leap back and forth between orbits.

Although the various stationary orbits of electrons in the hydrogen atom can be characterized by definite energies, it turns out that a stationary orbit, which is also called a quantum state, is not uniquely characterized by energy. More information is needed to tell which orbit is which. Specifically, we need to know the electron's angular momentum, which is essentially the speed at which the electron revolves around the proton. In quantum theory the angular momentum of the electron in a hydrogen atom can assume only discrete values. These values are multiples of a well-defined, smallest-possible, nonzero angular momentum, which is usually given in terms of Planck's constant, namely, $h/2\pi$ (this quantity is usually denoted by $\hbar$). Thus, possible angular momenta are 0, $\hbar$, $2\hbar$, $3\hbar$, and so on.

Spin—A Peculiarity of Quantum Theory

Nearly fifty years ago physicists noted that the electron not only revolves about the proton but also rotates about its own axis, that is, it has its own angular momentum. If we think of an electron as a small ball rotating about an axis, the angular momentum of such an electron can of course assume arbitrary values. This is not the case in quantum theory, however, where one finds that the individual angular momentum of an electron is always $\frac{1}{2}\hbar$; that is, it is half as large as the smallest nonzero angular momentum value we observe in the motion of the electron around the proton in the hydrogen atom.

Angular momentum is a vector quantity (that is, it has direction as well as magnitude), and for this reason the angular momentum of an electron about its axis can assume either one of two discrete values: $+\frac{1}{2}\hbar$ or $-\frac{1}{2}\hbar$. This individual angular momentum of the electron is called its spin, and any electron can have spin $+\frac{1}{2}\hbar$ or $-\frac{1}{2}\hbar$. In other words, the electron can spin in either one of two opposite directions (figure 1.3).

The spin phenomenon exists only in quantum theory—there is nothing analogous to it in classical

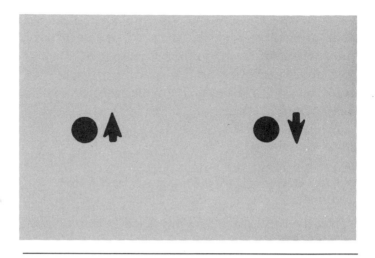

Figure 1.3
An electron has two possible spin states denoted by the arrows. The direction of the arrows is arbitrary. One might think that there are an infinite number of spin states since the spin vector can point in an infinite number of directions. However, in quantum theory this is not the case; only two different spin orientations are needed. The same is true for all particles with spin 1/2 (the electron, the positron, the proton, the neutron, and so on).

physics. The spin of an electron cannot be zero. Only two values are possible: $+\frac{1}{2}\hbar$ and $-\frac{1}{2}\hbar$. Again using the analogy of the electron as a small ball, we have no problem imagining a ball that does not rotate at all, in which case its spin is zero. In quantum theory, however, this is not allowed. The theory demands that electrons rotate about their axis and that their spin be either $+\frac{1}{2}\hbar$ or $-\frac{1}{2}\hbar$.*

The spin of an electron is an example of a quantum

* It is customary in elementary-particle physics to set the constant $\hbar$ equal to 1 and to measure all angular momenta, therefore, as multiples of $\hbar$. The $+\frac{1}{2}\hbar$ and $-\frac{1}{2}\hbar$ spins of the electron thus become $+\frac{1}{2}$ and $-\frac{1}{2}$.

number. An electron inside the hydrogen atom may be characterized by various quantum numbers, one for its energy, one for it angular momentum, and a third for its spin. The same holds true, of course, for more complicated atoms, for example, those of helium or uranium.

When the structure and especially the energy states of multiple-electron atoms were studied between 1910 and 1925, it was discovered that some rather unusual rules were obeyed. Toward the end of this period of intense research, twenty-five-year-old Austrian physicist Wolfgang Pauli found that most of these rules make sense only if no two electrons in an atom have the same quantum numbers (this of course is only relevant if an atom has more than one electron; that is, if it is more complex than a hydrogen atom). This rule is now called the Pauli exclusion principle. A few years after Pauli made this discovery, physicists discovered that the exclusion principle is a consequence of quantum mechanics combined with the theory of relativity. Furthermore, it was found that Pauli's exclusion principle is also valid for protons and neutrons, and in general for particles with nonintegral spin. However, it is not valid for particles with integral spin $(0,1,2,3 \ldots)$. Examples of such particles are pi mesons (hereafter referred to as π mesons or pions), which have spin 0 (in other words, π mesons have no individual angular momentum), and photons, the particles of light which have spin 1.

Positronium—A New System

Since it is our aim to discuss the structure of elementary particles within the framework of the quark model, it does not make much sense to engage here in an overly detailed discussion of the atom and atomic physics. However, a few words should be said about the structure of positronium, a special system that is very similar in structure to the hydrogen atom.

Positronium consists of an electron and its antiparticle, the positron (antiparticles will be discussed later), and is therefore an electrically bound state like the hydrogen atom since the electric charge of the positron equals the electric charge of the proton. This means that the electron and the positron attract each other just as the electron and the proton do. There is, however, one essential difference between positronium and the hydrogen atom: the proton is far heavier than the positron, which has a mass equal to the mass of the electron. (The mass of a particle always exactly equals the mass of its antiparticle.) For this reason alone we can no longer talk—at least not in the case of positronium—about an atomic nucleus, since electron and positron move around each other. We may think of positronium as a double-star system, in which two stars of the same size move around each other. The hydrogen atom, on the other hand, is comparable to a planetary system. A planet moves around the sun. Just as we can disregard the movement of the sun in astronomic considerations, since its mass is much larger than that of the planet, we can ignore the proton movement in the hydrogen atom.

Whereas the lifetime of a hydrogen atom is infinitely long, that of positronium is exceedingly brief. Positronium consists of the electron and its antiparticle; and all systems consisting of a particle and its antiparticle tend to annihilate themselves. Within moments of being generated, for example, in a laboratory, the positronium system decays into electromagnetic radiation. This decay is one of the most impressive illustrations of Einstein's theory of the direct conversion of mass (matter) into energy.

The annihilation of particles and antiparticles also occurs on other occasions, for instance, in the reaction of antiprotons with matter. In recent times it has become possible to generate intense antiproton beams— for example at the CERN laboratory in Geneva. When such beams collide with normal matter, the antiprotons and the protons in the nuclei of the target are annihilated, releasing large amounts of electromagnetic radiation, electrons, and positrons as well as electrically neutral particles called neutrinos.

Various laboratories have also succeeded in producing other forms of antimatter, such as the system consisting of an antiproton and an antineutron (called the antideuteron). In principle, it is also possible to create more complicated atomic nuclei, consisting of many antiprotons and antineutrons, if only sufficient amounts could be produced. This turns out to be exceedingly difficult, however, because the experimental apparatus is itself made of normal matter and the antiparticles have an unfortunate tendency to annihilate the matter of the apparatus that creates them. This makes the creation of antimatter a rather complicated business.

None of this, however, excludes the possibility that somewhere in the universe there exist vast systems—

stars, planets, perhaps entire galaxies—of antimatter. Astronomers are constantly searching for signs indicating the presence of antimatter in the universe. So far, however, this search has not been successful, and it appears that the universe does not contain large amounts of antimatter. In principle, however, the existence of antiiron and antigold is at least imaginable. Meteorites of antiiron colliding with the earth would annihilate upon impact, producing large amounts of energy, mostly in the form of electromagnetic radiation. It is a simple matter to calculate how much energy such a collision would produce. For example, the annihilation of 10 kg of antimatter and 10 kg of matter would produce sufficient energy to supply a state such as California with electricity for one year.

A poem by physicist Harold P. Furth alludes to a lecture by Dr. Edward Teller, (who played an important role in the construction of the first hydrogen bomb) on the subject of antimatter:

Perils of Modern Living
Harold P. Furth

Well up above the tropostrata
There is a region stark and stellar
Where, on a streak of anti-matter,
Lived Dr. Edward Anti-Teller.

Remote from Fusion's origin,
He lived unguessed and unawares
With all his antikith and kin,
And kept macassars on his chairs.

One morning, idling by the sea,
He spied a tin of monstrous girth
That bore three letters: A. E. C.
Out stepped a visitor from Earth.

A Look Inside the Atom

> Then, shouting gladly o'er the sands,
> Met two who in their alien ways
> Were like as lentils. Their right hands
> Clasped, and the rest was gamma rays.

Published in 1955, the poem was already outdated one year later, when it was determined that the transition from matter to antimatter should also entail a mirroring effect in space. Strictly speaking, Dr. Teller would have to hold out his right hand and Dr. Anti-Teller his left.

It is useful to take a somewhat closer look at the structure of positronium. Both the positron and the electron have spin, and the motion of the electron and the positron around each other can be described in terms of the angular momentum they have relative to

---

Figure 1.4

Parapositronium and orthopositronium differ from each other in spin orientation. In parapositronium (a) the spins are opposite each other. In orthopositronium (b) both spins point in the same direction.

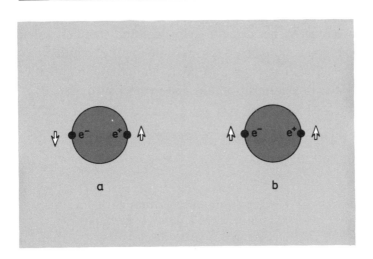

each other. The spins of these two particles can point in any direction—it is only the relative orientation of the two spins that counts. It is quickly apparent that positronium has only two possible spin states (figure 1.4): either the two spins are in the same direction or they are in opposite directions. If the spins are in the same direction, the state is called orthopositronium; if they are opposite, we speak of parapositronium. These two states can be easily distinguished from each other in the laboratory, and it turns out that parapositronium has a much longer lifetime than orthopositronium. We shall see later that this is because the electron and positron in positronium annihilate each other in an electromagnetic process.

Having introduced ortho- and parapositronium into the discussion, we should really speak of many possible quantum states of positronium because the ortho configuration and the para configuration can have very different angular momenta. Also, the two constituents of positronium can be either very far away from each other or very close to each other. To adequately characterize a positronium state, therefore, we require, besides the spin quantum number, other quantum numbers, such as that for the angular momentum. There are, in fact, an infinite number of states of parapositronium and orthopositronium.

II

The Unified Theory
of Electricity and
Magnetism

Today nearly everyone is familiar with electromagnetic interactions, for they are what is responsible for the existence of electric currents, and without them no television set, no car would work, However, we have found out in the course of the past eighty years that electromagnetic interactions are also responsible for the structure of atoms, the binding of atoms into molecules, the formation of crystals, and so on.

Today we subsume all these different phenomena into one theory—the theory of electrodynamics—but in 1800 the situation was quite different. At that time there were three classes of phenomena known to physicists: the attraction or repulsion of charged ob-

jects, magnetic phenomena, and light phenomena—
all three being merely different manifestations of the
same underlying interaction. At the time, however, no
one suspected that these three phenomena were intimately related. The situation did not change until
1820, when the Danish physicist Hans Christian Oersted found by accident that electric currents can produce magnetic forces, which, for example, can alter
the direction of a compass needle.

Eleven years after Oersted's discovery, the English
physicist Michael Faraday found that sudden changes
in magnetic fields can produce electric currents, a
phenomenon used on a large scale in electric power
plants. Although Faraday, upon first hearing of Oersted's principle years earlier, had immediately
guessed that the converse must be true, it took him
nine years of experimental work to confirm what he
had guessed.

In those days, physicists thought of electric and
gravitational forces as acting over large distances between different physical objects. For example, it was
thought that the sun attracts the earth because of the
sun's gravitational force. Specifically, it was held that
the sun's gravitational force acts directly on the earth
and has nothing to do with the space separating the
earth from the sun and the space surrounding the
earth. After many years of experimental work, Faraday decided to abandon this concept of "action at a
distance," at least for electrical forces. (Later Einstein made the same concession with regard to gravitational forces.) Faraday thought that electric forces
(the attraction or repulsion of electrically charged objects) were caused by lines of force that emanated
from the charged objects and filled the space between

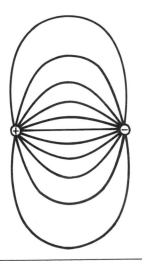

Figure 2.1
The lines of force connecting a negatively and a positively charged object.

the objects. In other words, electrically charged objects attract or repel each other because the space around them has a special property: it is filled with electrical lines of force (figure 2.1).

With this idea the modern notion of the electromagnetic field was born. As it turns out, the concept of the electromagnetic field—or, more specifically, the concept of field—is perhaps *the* most fundamental concept of physics. All modern theories of the elementary particles deal exclusively with fields.

At first Faraday thought of electromagnetic fields only in association with electrically charged objects. However, it was not long before he realized that such fields do not necessarily have to originate from charged objects but can exist independently. Furthermore, he felt that light might be nothing more than an electromagnetic phenomenon.

33

Maxwell's Equations

Faraday did not succeed in constructing a complete theory on the basis of his rather intuitive ideas. This very difficult task was left to the Scottish physicist James Maxwell (figure 2.2), who in 1861 derived the equations that describe electromagnetic interactions. If we look at these equations today, more than a century later, we see that they are a total success. They are still in use in essentially their original form, even after those two great discoveries that occurred after 1861: relativity and quantum mechanics.

Today we know that Maxwell's equations correctly

Figure 2.2

The Scottish physicist James Clerk Maxwell, who formulated the correct theory of electromagnetic phenomena, one of the greatest intellectual achievements of all times.

describe all electromagnetic phenomena, from the huge electromagnetic fields of the galaxies to those occurring within a space on the minute order of 10^{-16} cm, $1/10\,000$ of the diameter of an atomic nucleus.

What has changed in the years since 1861, however, is the interpretation of Maxwell's equations. Using the modern language of quantum field theory, we say that Maxwell's equations describe the propagation in space of electromagnetic quanta called photons (the constituent particles of lights). Even the electric attraction or repulsion between charged objects can be described in this manner. For example, two electrons flying through space influence each other by means of their electric repulsion (figure 2.3). (One can also speak of the scattering of electrons in

---

Figure 2.3

The repulsion of two electrons. The two electrons exchange photon quanta (called *virtual* photon quanta to distinguish them from "real" photons), and this exchange is what causes the electromagnetic repulsion between the two.

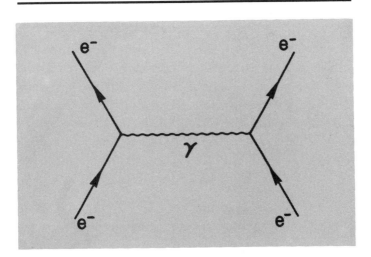

Figure 2.4

The American physicist Richard Feynman, one of the founders of the theory of quantum electrodynamics. Feynman also made important contributions to the physics of elementary particles. He received the Nobel Prize for physics in 1964.

an instance such as this.) Pictures like figure 2.3, which were invented by Richard Feynman (figure 2.4), are called Feynman diagrams. They are very useful in describing the electromagnetic processes in quantum mechanics.

Virtual Photons

According to quantum theory, a light beam consists of many photon quanta. Photons are particles with no rest mass; since their mass is zero, they move at the speed of light. The amount of energy carried by the photons is determined by the wavelength of the light. The shorter the wavelength, the greater the energy. For example, photons of red light have less energy than photons of blue light, since the wavelength of red light is longer than the wavelength of blue light.

"Real" photons, like electrons, have an infinite life, provided they do not interact with other particles. The life-span of what are called *virtual* photons, on the other hand, is very brief. Virtual photons are a consequence of the uncertainty principle. The interaction of two electrons by means of the exchange of a virtual photon occurs in a very short time, and a photon whose existence is so brief need not be massless. According to Heisenberg's uncertainty principle, it can have a mass, and as a matter of fact, the briefer the existence, the larger its mass can be. For example, modern high-energy accelerators can produce virtual photons with a mass of more than 30 GeV, that is, more than thirty times the mass of the proton.

The strength of an interaction in quantum physics is usually described by a strength parameter called the coupling constant. If this constant is greater than unity, the interaction is considered strong; if it is much less than unity, the interaction is weak. The electromagnetic interaction is described by a strength parameter called the fine structure constant. This coupling constant was originally introduced by the German physicist Arnold Sommerfeld early in this century. It is a pure number and had to be determined by experiment: $\alpha = 1/137.036 = 0.0073$.

The fact that α is such a small number is very useful. It allows us to calculate to a very high degree of accuracy, the quantum corrections for electromagnetic phenomena. So far, the results of tests to verify the theory of quantum electrodynamics are in splendid agreement with theoretical calculations.

Dirac's Equation and Antimatter

In 1928, while trying to combine the newly developed quantum theory with the ideas of relativity and Maxwell's equations, the English physicist Paul Dirac derived an equation for the propagation of charged particles which had surprising properties. Along with correctly describing the physical properties of electrons, Dirac's equation described the physical properties of a new particle, one having the same mass as the electron but which is positively charged. Shortly after Dirac obtained his result, the new particle, called the positron—the antiparticle of the electron— was observed in cosmic ray experiments.

Dirac concluded that all particles described by his equation must have an antiparticle, just as the electron has its positron, and today we know that this holds true for all subatomic particles. For example, there is an antiproton with an electric charge of -1, which was discovered in the 1950s.

The Long Life of Orthopositronium

The difference in the lifetime of the two forms of positronium can now be explained more fully. In parapositronium, the electron and the positron anni-

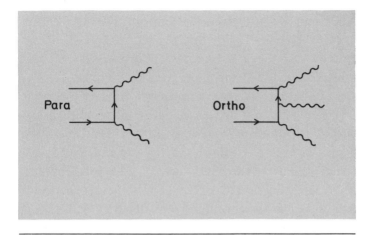

Figure 2.5
The annihilation of para- and orthopositronium into photons. The electron and positron enter from the left (straight lines). The photons are represented by wavy lines. Parapositronium decays into two photons, orthopositronium into three.

hilate each other and produce two photon quanta (figure 2.5). The annihilation of orthopositronium can be described in a similar fashion; in this instance, however, three photon quanta are produced. The fact that the annihilation of parapositronium leads to two photons and the annihilation of orthopositronium leads to three photons is a special consequence of Maxwell's equations, for two photons can never produce a state with angular momentum 1 (in units of $\hbar$). Since the two constituent particles in parapositronium have opposite spin, it has zero angular momentum (we are here considering the ground state of parapositronium, where the orbital angular momentum is zero). On the other hand, orthopositronium is the bound state, in which both constituents have the same spin and therefore the total angular momentum of the system is 1. For this reason, orthopositronium cannot disintegrate into two photons; the process must produce three or more photons.

In quantum electrodynamics we must pay a price for each photon that participates in a reaction between an electron and a positron. This payment comes in the form of α, the fine-structure coupling constant. Parapositronium decays into two photons. Therefore the square of the fine-structure constant will appear in the formula for its decay rate. The corresponding formula describing the decay of orthopositronium contains α to the third power because the decay produces three photons, and consequently the decay rate of orthopositronium is reduced by a factor $\alpha = 1/137$ compared to the decay rate of parapositronium. For this reason the lifetime of orthopositronium is more than one hundred times longer than that of parapositronium.

To summarize, the most important differences between electrons or positrons and photons are that electrons and positrons have an electric charge, a rest mass, and a spin of ½. The photon is electrically neutral, has no rest mass, and has spin 1 (this last is a special consequence of Maxwell's equations). The positron is the antiparticle of the electron. The photon is its own antiparticle. (Sometimes a particle is its own antiparticle. Of course, this can only be the case for neutral particles such as the photon.)

III

The Strong Interaction

When physicists first considered the structure of the atomic nucleus, they confronted a mystery. They knew that the nucleus was composed of nucleons— the protons and neutrons—and that heavy elements have many protons in the nucleus (uranium, for example, has ninety-two). But how can such a large quantity of protons be reconciled with the laws of electrodynamics, which prohibit the existence of so many particles of like charge in such a tiny space? The electric repulsion between these protons is so great that the nucleus should explode. According to the laws of electrodynamics, therefore, atomic nuclei should not be stable. There is only one solution to the problem: there must exist in nature another basic force that keeps the atomic nucleus together. This force indeed exists and is called the strong nuclear force or the strong interaction. It is the force responsible for the formation of atomic nuclei. It is very strong indeed: a naive estimate is that the strong interaction is at least one hundred times stronger than the electromagnetic interaction, whose strength is given by the fine-structure constant α.

Of course we cannot help but ask at once why we do not observe the effects of the strong interaction in everyday life the way we observe electromagnetic effects. The answer lies in a special property of the strong interaction: it is active only at very short distances, distances on the order of 10^{-13} cm. As soon as nucleons are removed farther apart from each other than 10^{-13} cm, the strong interaction has scarcely any effect. The dominant force between nucleons at such distances is the electromagnetic force. However, if the nucleons are brought close to each other again, the strong interaction very rapidly dominates. This is why atomic nuclei are stable and do not explode. Nucleons inside a nucleus are indeed very close to each other (10^{-13} cm or less), and their behavior is dictated by the strong nuclear force, which dominates over the electric repulsion between the protons.

The strong interaction discriminates between nucleons and the electrons in the atomic cloud. The electrons are not affected by the strong interaction at all. The only interaction of relevance to the atomic cloud is electromagnetic interaction. We shall see later why this is so.

Although we cannot see the effects of the strong interaction directly in everyday life, we can at least observe them indirectly. For example, the strong interaction is exceedingly important in nuclear power plants; the energy produced by a nuclear reactor is

The Strong Interaction

obtained through rearrangements of nucleons. It is this rearrangement that produces energy. The same principle applies in atomic and hydrogen bombs.

The Carrier of the Strong Force

We mentioned in chapter 2 that the interaction between charged objects (electric attraction or repulsion) is governed by the exchange of virtual photon quanta between them. The question immediately comes to mind as to whether we can understand the strong interaction between strongly interacting particles by means of a similar exchange principle. The suggestion that we may do so was made about fifty years ago by the Japanese physicist Hideki Yukawa, who proposed that the strong interaction results from the exchange between nucleons of quanta called mesons (figure 3.1).

Figure 3.1
The strong interaction (for example, scattering) between two nucleons via the exchange of virtual meson quanta.

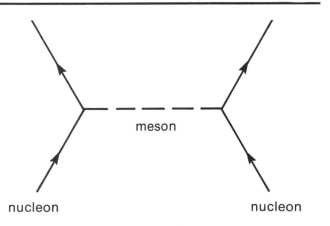

meson

nucleon   nucleon

It is a characteristic of interactions caused by the exchange of virtual quanta that the range of interaction is intimately related to the rest mass of the quanta exchanged. Photons, for example, are massless, and so the range of electromagnetic interaction for charged particles is infinite. This means that electric forces can be felt even at relatively large distances. There is no typical distance for the electromagnetic interaction as there is for the strong interaction. For example, we cannot say that electric forces can be neglected if we are more than 10 m away from a charged object, but we can say that the strong interaction is irrelevant once we are more than 10^{-13} cm away from a strongly interacting particle, such as a proton. Using this property of the strong interaction, Yukawa determined the mass of the hypothetical mesons that were supposed to produce the strong interaction. He came up with a mass of about 100 MeV, that is, about one-tenth the mass of the proton.

It took more than fifteen years for the existence of mesons to be firmly established by experiment. They are called π mesons (or pions) and have a mass of 140 MeV. There are three of them: π^+, π^-, and the π^0 (which is electrically neutral). The π^- particle is the antiparticle of the π^+; the π^0 particle is its own antiparticle.

The Art of Perturbation Theory

Except for the obvious difference that π mesons have mass and photons are massless, the introduction of mesons as the quanta of the strong interaction al-

The Strong Interaction

lows us to develop a theory of the strong interaction that is quite similar to the theory of electrodynamics. Still, we at once encounter one serious difficulty. Suppose we describe the strength of the strong interaction in a manner analogous to electrodynamic interaction and introduce a strength parameter analogous to the fine-structure constant α. It turns out that this constant is rather large—on the order of 10. Unfortunately, ascribing such a large value to the strong interaction strength parameter makes it impossible to develop a real theory of the strong interaction. The argument goes like this. Two electrons interact via the exchange of one virtual photon quantum between them. It is possible, though, for two photons to be exchanged instead of one, but the probability of this happening is very small, on the order of α^2, which is less than 1 in 10 000. Therefore we can neglect the two-photon process unless we need to be utterly precise. This method of calculation is called perturbation theory.

The smallness of the fine-structure constant allows us to make very precise calculations in quantum electrodynamics. In the case of the strong interaction, however, it makes little sense to use perturbation theory since the corresponding strength parameter is greater than unity (10 in our case). Let us look at the interaction between two nucleons as described by the exchange of π mesons. Consider the possibility that two π mesons are exchanged instead of one. The probability of such an occurrence is determined by the strength parameter squared, which in our case is 100. This means that there indeed exists a high probability that two rather than one meson are exchanged. Thus we encounter a situation that differs considerably from the one in electrodynamics, and perturbation theory makes no sense at all. For this reason, little progress was made in the past toward understanding the strong interaction. One may, of course, think that this failure is simply due to the fact that we do not know how to make the right calculations in the Yukawa theory (no one, so far, has found a method besides perturbation theory to perform the calculations). It appears, though, that this is not the reason for our lack of progress. In recent years we have learned much that is new about the strong interaction. The study of the behavior of nucleons under highly unusual circumstances (for example, in high-energy collisions with electrons) has provided invaluable insight into the dynamics of the strong interaction. For about the past eleven years physicists have been working on a special theory of the strong interaction, called quantum chromodynamics (QCD), and for the first time in the history of particle physics there is hope of understanding all strong interaction phenomena within a theory remarkably similar to Maxwell's theory of electrodynamics. This new theory is based on the notion of quarks, and it is one of the aims of this book to introduce readers to the ideas of quarks and chromodynamics.

IV

How Many Elementary Particles?

Around 1932 physicists were aware of four different particles as constituents of normal matter or of electromagnetic radiation: electrons, protons, neutrons, and photons. It was generally assumed that these particles were elementary; that is, they were not composed of smaller parts. Today, however, we know of more than one hundred such elementary particles, and it has become clear that most of them, including the proton and the neutron, are not elementary at all.

The most peculiar particle in this plethora of particles is without doubt the neutrino, whose existence was predicted in 1932 by Pauli (then professor of theoretical physics at Zurich). Pauli deduced the existence of the neutrino from a phenomenon observed in the decay of unstable nuclei called beta decay (hereafter referred to as β decay). In 1896 it was discovered that certain atoms decay into other atoms (later it turned out that it was the atomic nuclei which were unstable). Sometimes these decay processes are accompanied by the emission of electrons, in which case the process is called β decay. This decay is caused by an interaction called the weak interaction, which we shall discuss in detail later.

The decay of the neutron, which we have already discussed, is an example of β decay. The neutron decays into a proton by emitting an electron. Recall that the neutron is about 1.3 MeV heavier than the proton. If the neutron were lighter than the proton, matters would be reversed—the proton would decay into a neutron by emitting a positron. This is not the case. If it were, hydrogen would be unstable and complex organic matter and life could not exist.

It should be mentioned at this point that we do not understand why the neutron is heavier than the proton. Indeed, an unbiased physicist would have to assume the opposite, by the following logic. It is reasonable to think that the difference in mass between the proton and the neutron is related to electromagnetic interaction since the proton has an electric field and the neutron does not. If we rob the proton of its charge, we would expect the neutron and proton to have the same mass. The proton is therefore logically expected to be heavier than the neutron by an amount corresponding to the energy needed to create the electric field around it.

We notice something analogous in pions (mesons). The positively charged pions (the π^+ mesons) are

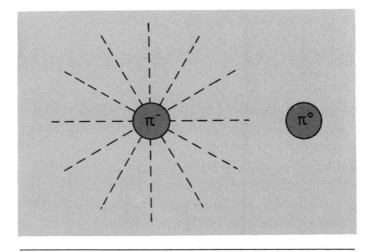

Figure 4.1
Charged pions (π^+ mesons), unlike neutral pions, are surrounded by an electric field. The energy of the electric field surrounding a charged pion corresponds to the difference between the charged and neutral pion masses. Without electromagnetic interaction, the charged and neutral pion would have exactly the same mass.

slightly heavier than the neutral ones and the energy of the electromagnetic field surrounding these particles can be calculated precisely. The difference between the mass of the charged pion and that of the neutral pion corresponds exactly to the energy of the electromagnetic field. We conclude, then, that the charged pions are heavier than the neutral ones because they have an electromagnetic field (figure 4.1).

Oddly enough, the situation is reversed with nucleons: the neutral particle is heavier than the charged one. Why this is so we still do not know.

Energy Crisis in Physics

One of the fundamental laws of physics is the law of conservation of energy, which states that the total energy in all physical processes is conserved. Provided a reaction takes place in an isolated system, where no energy is lost to or supplied from the outside, the energy at the start equals the energy at the finish. The law of conservation of energy is also valid for processes involving elementary particles and specifically for the energy released in the neutron decay process. The sum of the energies of the final particles—proton and electron—should be equal to the energy of the neutron (the initial energy of the neutron at rest is given by its rest mass).

Experiments on β decay provided surprising results: the sum of the energies of the final particles was less than the initial energy. An energy crisis arose. Some physicists, including Bohr, the father of atomic physics, even considered the possibility that the law of conservation of energy could be violated, at least in such rare instances as this.

Pauli finally solved the problem by proposing the existence of another neutral particle, one that is emitted with the electron in the decay process. The Italian physicist Enrico Fermi later named the particle neutrino ("tiny neutron"). However, it turns out to be useful to think of the neutral particle emitted in the neutron decay process as an antineutrino or, more precisely, an electron-antineutrino, denoted by the symbol $\bar{\nu}_e$. (We shall see later that there are other neutrinos.) We can now write a formula for neutron

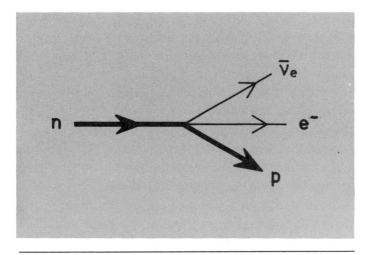

Figure 4.2
Neutron decay. The incoming neutron disintegrates by emitting an electron e⁻, a proton p, and an electron-antineutrino $\bar{\nu}_e$.

decay (figure 4.2) that includes antineutrino emission:

$$n \rightarrow p + e^- + \bar{\nu}_e$$

The antineutrino therefore accounts for the missing energy in neutron decay, and the day is saved for the principle of energy conservation, though at the expense of the introduction of a new particle.

Pauli himself was less than delighted by his discovery. He believed that the new particle could never be observed directly and that his hypothesis would therefore remain just that. As it turned out, Pauli did not underestimate the difficulties of finding the neutrino, for it took more than twenty years for physicists to verify his hypothesis. Not until the beginning of the 1950s was the neutrino finally discovered through nuclear reactors, which produce neutrinos in large quantities. (A large fraction of the energy generated inside a nuclear reactor is in the form of neutrino radiation.) Today we know far more about neutrinos, and various research centers can produce intense beams of neutrinos for research purposes. Pauli would be quite surprised to hear that his elusive particle is being "used" in modern high-energy physics research.

Neutrinos are very light particles. Their mass is much smaller than the electron mass. In fact, it may be that they are massless, like photons. Various experiments have set an upper limit on the mass of the electron-antineutrino emitted in the neutron decay process. It is on the order of 30 eV; that is, the neutrino's mass must be 10 000 times less than that of the lightest particle we know, the electron.

In 1979 physicists in Moscow conducted an experiment to lower the limit on the neutrino mass even further and found indications that the neutrino is not massless but has a mass between 10 and 30 eV. However, the findings of the Russian physicists have been challenged, and the situation remains unclear. It will probably take years to determine whether the neutrino is a massive particle or not.

The issue of a nonvanishing neutrino mass is highly important in other fields of physics, especially astrophysics. If neutrinos are massive and have a mass of more than a few electronvolts, a large part of the total mass of the universe would consist of neutrinos.

The physical characteristics of the neutrino are remarkable. It ignores the strong and electromagnetic interactions and is affected only by the weak interaction, the same interaction responsible for its birth in the neutron decay process. This implies that neutrinos interact only very weakly with normal matter and that they can penetrate large amounts of matter with ease. For example, a beam of neutrinos can pass through the earth, or even through the sun, without being much disturbed and only rarely interacting with other particles.

However, these rare interactions are of great interest to physicists. The study of high-energy interactions of neutrinos with atomic nuclei, carried out particularly at CERN and at FNAL since 1972, have provided numerous insights into the structure of matter, some of which we shall discuss later.

The properties of the neutrino are portrayed in the following poem by John Updike, written in 1960:

Cosmic Gall
John Updike

Neutrinos, they are very small.
    They have no charge and have no mass
And do not interact at all.
The earth is just a silly ball
    To them, through which they simply pass,
Like dustmaids down a drafty hall
    Or photons through a sheet of glass.
    They snub the most exquisite gas,
Ignore the most substantial wall,
    Cold-shoulder steel and sounding brass,
Insult the stallion in his stall,
    And, scorning barriers of class,
Infiltrate you and me! Like tall
And painless guillotines, they fall
    Down through our heads into the grass.
At night, they enter at Nepal
    And pierce the lover and his lass
From underneath the bed—you call
    It wonderful; I call it crass.

The Particle Zoo

Many other particles have been found since 1950, all of which are unstable and have very short lifetimes. They are produced in the collision of stable particles, for example, in proton-proton collisions, and within tiny fractions of a second have again turned into familiar particles of one kind or another. A detailed description of all these particles is beyond the framework of this book, and therefore we shall limit our discussion to a general classification of the particles.

It is useful to classify all these particles into two groups. The first group includes all strongly interacting particles, of which we already know five: the proton, the neutron, and the three π mesons. The strongly interacting particles are filed under the rubric *hadrons*. A variety of hadrons have been discovered either in cosmic rays or in accelerator laboratories, and it turns out that there are two classes of hadrons, the mesons and the baryons. Typical representatives of the meson class are the π mesons, the lightest known mesons. Other examples are the K mesons, new mesons whose mass is about one-half that of the proton; and the rho mesons (hereafter referred to as ρ mesons), whose mass is approximately three-fourths that of the proton.

It is characteristic of all mesons that they finally decay into electrons and positrons, neutrinos and photons. Nothing else remains. All mesons also have integral spin in units of $\hbar$. We already know that the π meson has spin 0; so does the K meson. The spin of the ρ meson is unity. However, there also are mesons of spin 2, 3, and so on. It is characteristic of mesons that in general their mass increases with their spin. Although experiments are not totally unequivocal as yet, particle physicists today believe that there are an infinite number of mesons, with variably high spin. Recall that, according to the laws of quantum mechanics, the spin of particles can assume only integer or half-integer values of $\hbar$.)

The second class of hadrons are the baryons, which include protons, neutrons, and many other particles. To mention only a few, there are hyperons (the lambda, sigma, and xi particles, hereafter referred to as Λ, Σ, and Ξ particles) and the delta particles (hereafter referred to as Δ particles). All these baryons are heavier than the proton (hence the name *baryon,* which means *heavy* in Greek). The spin of the baryons is always a noninteger, that is, $1/2$, $3/2$, and so on. For example, the spin of the nucleons is $1/2$, and the spin of the Δ particles is $3/2$.

All baryons with the exception of the proton (and there, too, we have our doubts, as mentioned at the end of the book) finally decay, so that the only strongly interacting particle that remains is the proton; all other remaining particles do not interact strongly. For example, the lambda hyperon (mass 1116 MeV) decays within 10^{-10} s to a nucleon and a pion (for example, a proton and a π^-, figure 4.3). Then the pion decays so that all we have left at the end are a few neutrinos and an electron plus a proton. (The decay of the pion will be discussed in more detail later.)

The fact that baryons finally decay into protons and other particles that lack the capacity to interact strongly is a consequence of an important conservation law, the law of conservation of baryon number.

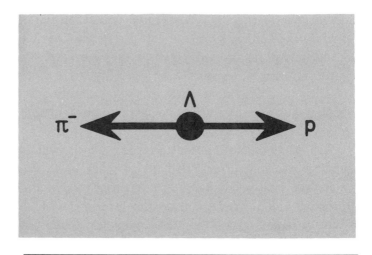

Figure 4.3
A Λ hyperon at rest decays to a proton and a positively charged pion. Because of momentum conservation, the two particles shoot off in opposite directions. The momentum of the initial particle is zero since it is at rest. The momentum of the proton must be equal in magnitude to that of the pion but opposite in sign.

All elementary-particle interactions familiar to us obey a conservation law that states that the total number of baryons minus the total number of anti-baryons is conserved during the interaction. The law can be illustrated as follows. We ascribe a number to each hadron, called the baryon number A. By definition, the proton and neutron have $A = 1$ and the anti-proton and antineutron have $A = -1$. Mesons have $A = 0$. The conservation of baryon number means that the sum of all these numbers, the total baryon number, remains unchanged in any interaction. For example, in neutron decay the initial baryon number is $+1$. The decay products also have baryon number $+1$, which is given by the baryon number of the proton. The electron and neutrino have baryon number 0. If we consider the collision of two protons at high energy, where typically a large number of new particles is produced, the initial baryon number is 2. Therefore, the baryon number of the final state, consisting of many particles, should be 2 as well. This can happen any number of ways, the simplest being to end up with two protons or, more generally, two nucleons and any number of mesons or other particles having a baryon number of 0. However, it is also possible to end up with several baryons and antibaryons, for example,

$$p + p \rightarrow p + n + p + \bar{n}$$

The total baryon number of the final system is again 2 (the baryon number of the antineutron is -1).

In physics the law of conservation of baryon number is just as important as the law of conservation of electric charge. If the baryon number is indeed conserved, the proton is absolutely stable. Without this conservation law the proton could decay—into a positron and a neutral pion. There is nothing in physics to prevent such a decay except the law of conservation of baryon number. Therefore the stability of nuclear matter is attributed directly to this law.

How Stable Is the Proton?

Though to all appearances the proton seems absolutely stable, perhaps it does decay after all, perhaps over an immensely long span of time. Physicists have designed experiments to search for possible proton decay but have found none so far; all they can say with any certainty is that the proton has a life-span of at least 10^{30} years. This lower limit on the life-span of the proton is greater than the estimated age of the universe (roughly 10^{10} years). How, then, do physicists come up with such a result? The point is that when physicists look for proton decay they do not look at one individual proton, but instead investigate a macroscopic piece of matter which contains, for example, 10^{30} protons. Each of these protons may of course decay, and even if the proton life-span is exceedingly long (longer than the age of the universe), chances are that one of them will decay while the experiment is going on. It is in this manner that we arrive at the limit mentioned above.

Still questions remain. Is the proton absolutely stable or is its life-span finite? Why worry so much about the question of proton stability? Why not simply assume that baryon number is strictly conserved and that the proton is absolutely stable?

There is something slightly odd about an absolutely conserved baryon number, and today many experts are convinced that the proton can decay, albeit within a time span of more than 10^{30} years. What is so unsatisfactory about an absolutely conserved baryon number? There is an important difference between baryon number and electric charge. Electric charge is not merely a number assigned to the various elementary particles (-1 to an electron, 0 to the neutrino, and so forth) and conserved in elementary processes; electric charge also determines the dynamic behavior of a particle. Specifically, a charged particle has an electromagnetic field around itself, which extends infinitely and can be measured even at considerable distances from the particle. Moreover, we can measure an object's electric charge without looking at the object directly, just by measuring the electric field that surrounds it. This is not the case with baryon number. A particle with a nonzero baryon number does not create a field analogous to an electromagnetic field. The baryon number is simply a bookkeeping device that tells us that the number of baryons in a given volume remains unchanged and that no baryons are added or subtracted. The baryon number—or baryonic charge, as it is sometimes called—has no dynamic significance, as the electric charge, of course, does.

Readers, however, might wonder whether there does not perhaps exist some kind of field that is associated with the baryon number but is too weak to be easily observed, much weaker, perhaps, than the electromagnetic field surrounding a charged object. How weak would such a field have to be to have escaped detection? Let us go through a simple thought experiment. We suppose that there is an analog to the electromagnetic field, which we shall call the baryonic field, and that this field is tied as strongly to the baryonic charge as the electromagnetic field is to the electric charge. We know that normal matter consists of baryons and not of antibaryons. Consequently, each macroscopic piece of matter must have a very large baryonic charge. Specifically, there would exist very strong baryonic repulsion between such large bodies of matter as the sun and the earth, and we can easily see how strong this repulsion must be compared to the gravitational attractive force between these two bodies. The result of such a comparison is enormous: the baryonic field would have to be more than 10^{36} times stronger than the gravitational attraction! Indeed, this absurd result provides us with a convincing argument against the existence of any such baryonic field.

We have already mentioned another peculiar difference between electric and baryonic charges. In general, pieces of macroscopic matter have no charge; they are neutral. However this is not so with respect to the baryonic charge. (An automobile, for example, has a positive baryon number of about 10^{30}.) We find no evidence, however, for the existence of large quantities of antibaryons despite the fact that we can produce them quite copiously in the laboratory. It appears as if Nature has some profound objection to antibaryons. All the stars in our galaxy consist of baryons and not of antibaryons. This is peculiar, for in the realm of elementary particles there exists a symmetry of baryons and antibaryons. There appears to be nothing to differentiate between baryons and antibaryons when baryon number is conserved absolutely, and therefore one would expect the universe to contain as many baryons as antibaryons. However the nonexistence of any significant amounts of antimatter in the universe becomes comprehensible if we relinquish the law of conservation of baryon number, thus smoothing the way for the decay of the proton.

Later we shall see that modern theories of elementary particles do not automatically imply the conservation of baryon number. As a matter of fact, in some theories, which we shall discuss, the proton has a finite lifetime.

How Many Elementary Particles?

Leptons

Besides the mesons and baryons (the strongly interacting particles known collectively as hadrons), there is another group of particles we should mention here. In general, all particles that do not participate in the strong interaction but do have spin ½ are called leptons. Electrons and neutrinos are leptons, for instance, but photons are not since they have spin 1.

During the last twenty years, many new hadrons have been discovered so that today we know of several hundreds of them. On the other hand, the number of leptons is fairly limited. Until recently only four leptons were known; besides the electron and neutrino, scientists have discovered another charged lepton, the muon, also denoted by μ, and a second neutrino, the myonneutrino (denoted by v_μ). The muon is roughly 200 times heavier than the electron and disintegrates within $1/1,000,000$ of a second into an electron, an electron-antineutrino, and a myonneutrino.

The word *lepton* (from the Greek *leptos* for *small, fine, light in weight*) is indeed quite appropriate for the four known leptons—e^-, v_e, μ^-, v_μ—are exceedingly light compared to the strongly interacting particles. In fact, most physicists speculated until recently that leptons had a small mass since they do not participate in the strong interactions. However, in 1975 experimentalists at SLAC observed certain effects that were incomprehensible without the existence of another charged lepton, this time a heavy (!) lepton with a mass roughly twice that of the proton. Subsequent experiments at SLAC and DESY have confirmed this hunch, and the new lepton (named tau, hereafter referred to as the τ lepton) is now a firmly established particle.

Its lifetime, however, is brief—about 10^{-13} s—and it decays via the weak interaction process similar to muon decay. Having discovered this relatively massive new τ lepton, scientists had little choice but to invent a new class of leptons, the heavy leptons, and to appoint the newly discovered τ lepton as its first member. Of course, the name *heavy lepton* is a contradiction and shows once again how illogical expressions used in physics or in science in general can sometimes be.

The charged leptons participate in both electromagnetic and weak interactions, the neutral leptons (neutrinos) only in the weak interaction. As of now, we know of two neutrinos, the electron-antineutrino from β decay and the muon neutrino, which have both been observed directly. In 1978 indirect confirmation from various experiments suggested that the τ lepton, like the muon, is associated with a new neutrino, the τ neutrino (hereafter denoted by ν_τ). Thus we can classify the leptons in the following three groups:

$$\begin{pmatrix} \nu_e \\ e^- \end{pmatrix} \quad \begin{pmatrix} \nu_\mu \\ \mu^- \end{pmatrix} \quad \begin{pmatrix} \nu_\tau \\ \tau^- \end{pmatrix}$$

We shall see later that the pairing of leptons has a deeper significance: it implies very special qualities of the leptons with regard to the weak interaction.

Quarks—Finally

If we compare the world of hadrons with the world of leptons, we see that there are an abundance of the former but only six leptons observed so far. This state of affairs corroborates our suspicion that leptons in some sense are more elementary than hadrons and that hadrons are composite systems, consisting of yet smaller and more elementary units. During the last fifteen years, we have discovered that this supposition is indeed correct, that the hadrons are made up of simpler particles, the quarks. This new concept of quarks was originally developed by Murray Gell-Mann and George Zweig of the California Institute of Technology (figure 4.4), and since its conception in 1964 the quark model of the hadrons has developed

Figure 4.4
The American physicists Murray Gell-Mann (left) and George Zweig (right). Gell-Mann's research has had a decisive impact on the development of theoretical physics since 1954. He received the Nobel Prize in physics for his work on the classification of elementary particles in 1969. He holds the Millikan chair of theoretical physics at the California Institute of Technology in Pasadena. Zweig is professor of physics at the same institution. Besides his work in particle physics, Zweig has also researched problems in biophysics.

from a rather bold hypothesis into a viable theory.

We shall see that the number 3 plays an especially significant role in quark theory. For example, a proton consists of three quarks, and it was the number three that led Gell-Mann to introduce the expression "quark." There is a passage in James Joyce's *Finnegans Wake* that reads:

> Three quarks for Muster Mark!
> Sure he hasn't got much of a bark
> And sure any he has it's all beside the mark.
> But O, Wreneagle Almighty, wouldn't un be a sky of a lark
> To see that old buzzard whooping about for uns shirt in the dark
> And he hunting round for uns speckled trousers around by Palmerstown Park?

Joyce's novel is full of plays on words that are difficult to understand. Many of them have never been deciphered. The novel describes the life of Mr. Finn, who sometimes appears as Mr. Mark. The "three quarks" denote the three children of Mr. Finn, by whom he himself is represented from time to time. Thus the association with particle physics becomes clear. The proton is associated with Mr. Finn; under certain circumstances the proton acts as if it consisted of three quarks.*

---

*In German, the word "quark" describes a special kind of soft cheese, but it is also used for "nonsense." Note that the title of this book is "Quarks," and not "Quark."

V

Mesons, Baryons, and Quarks

We mentioned earlier that there are far more hadrons than leptons and that this is one of the chief differences between these two kinds of particles. If we take a somewhat closer look at the physical properties of hadrons, however, we notice at once a number of regularities. One of these is that protons and neutrons, despite the difference in electric charge, are very similar to each other. Both have spin 1/2 and baryon number +1. Also, their masses are nearly equal—the neutron being only slightly heavier than the proton.

Isospin—A New Symmetry

Since both protons and neutrons participate in the strong interaction, Heisenberg suggested that this force affects both particles in the same way. This invariance in the effect of the strong interaction on neutrons and protons is called isospin invariance.

The concept of isospin symmetry represents an encouragingly symmetric aspect of nature, especially with respect to strong interactions. Isospin symmetry turns out not to be absolutely perfect, of course, since the electromagnetic interaction affects protons and neutrons differently and is therefore not symmetric. Also, the interaction of photons with protons is of course different from that of photons with neutrons. Neutrons interact with photons despite the vanishing neutron electric charge because the neutrons have a magnetic moment; that is, the motion of a neutron can be influenced by a magnetic field.

If the concept of isospin invariance is applicable to the world of hadrons, we would expect hadrons to be divided into various groups. All members of one of these groups would have the same spin, nearly the same mass, and the same baryon number but different electric charge. Looking at the properties of the hadrons, we immediately realize that this is indeed the case. Examples of such groups are the three π mesons—π^+, π^0, and π^-—and the so-called Δ particles. The latter were found in the early 1950s in the π nucleon scattering experiments and have a mass of 1232 MeV; that is, they are approximately 25 percent heavier than protons. There are four Δ particles—Δ^{++}, Δ^+, Δ^0, Δ^-—and of particular interest is the one with charge $+2$.

There are also particles that lack isospin partners. One such particle is the eta meson (hereafter referred to as the η meson), which is electrically neutral and about four times heavier than the π meson and exists all by itself.

Mesons, Baryons, and Quarks

In addition, we should mention the ρ mesons, which have spin 1 and a mass of about 770 MeV, and the neutral omega meson (hereafter referred to as the ω meson), which has spin 1 and a mass of 783 MeV. These particles are shown in figure 5.1. There are also many other particles, with even larger masses.

Figure 5.1
The masses of the light hadrons. The mass differences in the isospin multiplets are exaggerated slightly.

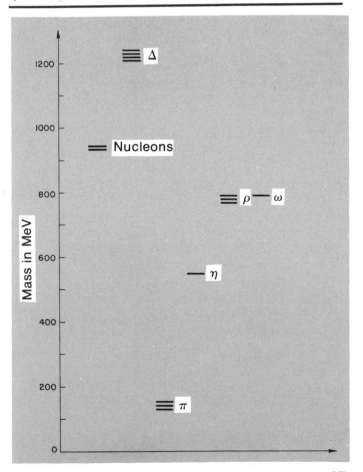

All of these particles can be grouped in isospin multiplets, similar to those shown in figure 5.1.

The isospin invariance of the strong interaction is an established fact of Nature. What, however, is the physical reason for isospin invariance? Today we know that hadrons are not elementary particles, but are made up of even smaller objects, the quarks. We suspect that isospin symmetry can be understood only by taking the quark structure into account.

First of all let us try to solve the following problem. We take another look at the isospin multiplets in figure 5.1 and try to construct a model in which these particles are made up of smaller particles. What physical properties should these smaller particles have? First of all, they must all have spin 1/2 because otherwise it would be impossible to construct spin 1/2 objects (nucleons, for example) from them. Furthermore, these constituent particles must have a nonvanishing baryon number since, for example, the proton has a baryon number of +1. In addition, they must have an electric charge because hadrons do.

Constructing mesons from spin 1/2 particles is a fairly easy matter. The same principle, after all, was used in atomic physics. The hydrogen atom, for example, is a system consisting of two spin 1/2 objects (proton and electron). Positronium, which consists of an electron and its own antiparticle, is even more closely related to a meson. Therefore, let us construct mesons from spin 1/2 particles and the corresponding antiparticles. We discover that such a system indeed has one of the properties of a meson: its spin is always integral (0,1,2...), like the spin of positronium.

Of course, the analogy between mesons and the hydrogen atom or positronium is useful only with respect to the spin. In other respects positronium behaves very differently from a meson. For example, positronium does not participate in the strong interaction.

Isospin and quarks

How many particles does one need to construct, for example, the meson multiplets in figure 5.1? Certainly more than one, since otherwise all mesons would be neutral and consist of a particle and its antiparticle. (Since the electric charges of a particle and its antiparticle are equal in magnitude but opposite in sign, the electric charge of a particle-antiparticle system is always zero.) Let us suppose we have two particles, which we call u and d quarks (up and down). With two types of quark (so-called quark flavors), we can form four different quark-antiquark systems, namely, $\bar{u}u$, $\bar{u}d$, $\bar{d}u$, and $\bar{d}d$. Also, we observe that the mesons always appear in groups of four (either π^+,π^0,π^-; η or ρ^+,ρ^-,ρ^0; ω). Therefore let us identify the mesons as $\bar{q}q$ systems.

What is the structure of the baryons? If we think of baryons as composite objects, they must consist of more than one quark. Obviously, baryons cannot be two-quark systems because the spin of a two-quark system cannot be a half integer (as the spin of the nucleon is). Therefore the simplest possibility for baryons is a three-quark system. Can we construct the baryons shown in figure 5.1 from our u and d quarks? Let us first consider the Δ particles. It is quite evident that there are only four ways of constructing qqq systems out of u and d quarks, namely uuu, uud, udd, and ddd. Thus we shall identify the four different Δ states with these four quark configurations, as follows:

$$\Delta^{++} = \text{uuu}$$
$$\Delta^{+} = \text{uud}$$
$$\Delta^{0} = \text{udd}$$
$$\Delta^{-} = \text{ddd}$$

Using this identification, we read off the quantum numbers of the quarks. What a surprise! The baryons are qqq systems. Therefore, the baryon number of a quark should be 1/3, that is, nonintegral. The electric charge of the u quark must be 1/3 of the charge of Δ^{++} (that is, 2/3). Similarly, the charge of the d quark must be $-1/3$. Thus for the first time in physics we come upon nonintegral electric charges.

We can also construct the nucleons out of u and d quarks. The proton, which has an electric charge, must have the quark structure uud. The neutron, which has no electric charge, has the structure ddu.

Let us return to the mesons. Can we reproduce the correct electric charges of the mesons by identifying them as $\bar{q}q$ states? Yes, it works: the charge of the $\bar{u}d$ system is -1, the charge of du is $+1$, and the charge of $\bar{u}u$ or $\bar{d}d$ is zero. Thus we identify

$$\pi^{+} = \bar{d}u$$
$$\pi^{-} = \bar{u}d$$
$$\rho^{+} = \bar{d}u$$
$$\rho^{-} = \bar{u}d$$

The neutral mesons π^{0}; η and ρ^{0}; ω are slightly more complicated because it turns out that they are superpositions of several neutral $\bar{q}q$ configurations. For the moment, however, let us not worry about this.

In general, we can state that the quark idea provides a very simple description of the various baryons and mesons. Mesons are $\bar{q}q$ systems, baryons are qqq systems. The mesons and baryons are shown schematically in figure 5.2.

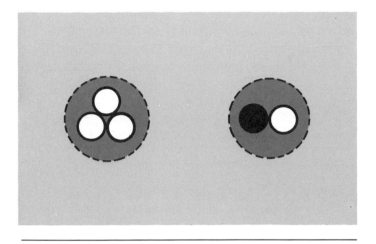

Figure 5.2
Mesons as quark-antiquark systems and baryons as three-quark systems.

Above we identified the π mesons and the ρ mesons with the same quark configuration. The nucleon and two of the Δ states also have a common quark configuration. These particles all have quite different masses however. The ρ meson is about 600 MeV heavier than the π meson, and the Δ particle is more than 300 MeV heavier than the nucleon. How, using the same combinations of quarks, can we describe particles with such different masses?

Luckily, the answer to this puzzle is very simple.

So far we have left out of our account that quarks have spin 1/2. The π mesons have spin 0, and so we have to arrange the spins of the quark and the antiquark in such a way that they cancel each other. Since ρ mesons have spin 1, the two spins of their quarks have to be aligned. In the nucleon, two of the three quark spins cancel each other so that the total spin is 1/2. The Δ particles have spin 3/2, which we obtain by aligning the spins of all three quarks.

Looking at the various spin arrangements in figure 5.3, we reach an important conclusion: the force be-

Figure 5.3
The quark spin structure of the mesons and baryons (of the π, ρ system and the nucleon, Δ) system.

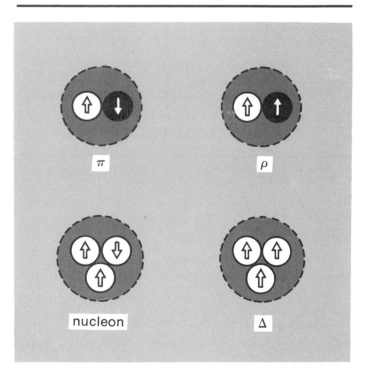

tween quarks must be such that the spins in a meson are opposed to each other. Also, it takes energy to align the spins in the same direction, as in a ρ meson, for example. The same holds true for baryons. It costs energy to align all the quark spins to arrive at a Δ particle. Any valid theory of the strong interaction based on quarks must incorporate and account for all these features.

Quarks—So Hard to Accept

Only a few people took the quark idea seriously when Gell-Mann and Zweig first proposed it in 1964. Believing in quarks seemed to require the acceptance of rather too many peculiarities, not only the unconventional electric charges of quarks but quite a few other mysterious features as well. One of these stumbling blocks was that hadrons always appear in the form $\bar{q}q$ or qqq. Why, for example, could there not be other combinations, qq states (called diquarks), maybe, or four-quark states (qqqq)? Experimenters on the lookout for such exotic hadrons have not found any to date. Instead they stubbornly keep observing $\bar{q}q$ or qqq hadrons.

Another problem with the quark concept is that quarks apparently do not exist as free particles, as electrons do, for example. Naively one might presume that the nucleon's quark structure is similar to the atom's nuclear structure, at least in the sense that a nucleus can be broken down into its constituent parts with relative ease. For the past fifteen years people have sought to do the same thing with nucleons, but without success. Today it is apparent that either free quarks do not exist at all (in other words, quarks are permanently bound inside the nucleons) or they are exceedingly heavy, at least fifteen times as heavy as the proton. This makes it very difficult to observe them in the laboratory. If free quarks existed, they would be rather freakish objects because of their non-integral electric charge. We should note, in particular, that at least one quark would be entirely stable as a result of the law of conservation of electric charge.

In the remaining chapters of this book, we shall see how all these problems can be resolved in a unique and simple theory of quark dynamics called quantum chromodynamics. Quarks in this theory are the fundamental particles of hadron physics, just as the electron is the fundamental particle of electrodynamics. Quarks indeed have all the peculiar qualities we have ascribed, but they will never have an opportunity to show these qualities directly because they are permanently bound inside the hadrons and can never be knocked free. Nevertheless, there are many ways of getting an indirect glimpse of quarks in nucleons. One of the best is described in chapter 6.

We should note right here, however, that it is still not certain that free quarks cannot exist. A group at Stanford University has been investigating various kinds of materials to find evidence for the existence of fractional charges, and it claims to have found particles carrying an electric charge of $1/3$. These particles may be free quarks, or they may be other as yet unspecified particles—we do not know. Many experts question the interpretation that the Stanford scientists gave their findings. Therefore it will at least be a few more years before we know whether particles with fractional charge exist or not.

We conclude this chapter with a quotation from Zweig that demonstrates the vehemence with which the theoretical physics community reacted to the quark model in 1964. Zweig writes in a recent Cal Tech report:

The reaction of the theoretical physics community to the model was generally not benign. Getting the CERN report published in the form that I wanted was so difficult that I finally gave up trying. When the physics department of a leading university was considering an appointment for me, their senior theorist, one of the most respected spokesmen for all of theoretical physics, blocked the appointment at a faculty meeting by passionately arguing that the model was the work of a "charlatan." The idea that hadrons, citizens of a nuclear democracy, were made of elementary particles with fractional quantum numbers did seem a bit rich. This idea, however, is apparently correct.

VI

The Proton Is X-rayed in California

We have introduced the concept of quarks as the constituent particles of hadrons in order to explain the various hadronic states observed in nature. Is there any possibility, though, of observing quarks more or less directly? To answer this question, let us recall how the structure of the atom was originally determined: not by tearing the atom apart but by a more subtle method. Rutherford used α particles to look inside the atom the way a physician uses x rays to look at a patient's organs. Perhaps something similar could be done with protons. This, at least, was the idea some people had in 1967.

Such a probe deep inside the proton was, of course, not to be accomplished by an experiment as simple as Rutherford's. Physicists could not just shoot α particles or high-energy nucleons at some targets. That

The Proton is X-rayed in California

would be the same as using hydrogen atoms instead of α particles in the original Rutherford experiment. Exploring the structure of the proton would require something considerably smaller, perhaps one of the leptons—an electron, for example. Since leptons cannot interact strongly, they are ideally suited for penetrating deep inside the nucleon.

For this purpose, it was decided that a huge accelerator should be built, one able to accelerate electrons to energies of 20 000 MeV, which is about twenty times the energy equivalent of the proton mass. What did physicists expect to happen once this accelerator

Figure 6.1
A look inside the SLAC machine. We see the two-mile-long vacuum pipe in which electrons are accelerated to a speed very close to the speed of light.

was built and electrons started smashing against the proton targets (figure 6.1)? They imagined that the electrons would penetrate the protons and be diverted by the nucleonic structure. Since electrons participate only in the electromagnetic interaction, the strongly interacting material inside the proton, whatever it is composed of, should not represent an obstacle for them. It is the distribution of electric charge inside the proton that determines whether or not a penetrating electron is diverted in its flight. Most physicists thought that the electric charge was distributed smoothly inside the proton. For this reason, they expected the electrons to scatter "softly" off the target protons, that is, only a small part of the energy and the momentum of the incoming electron would be transferred to the proton. It was thought that only rarely would hard scattering take place. Only rarely would the electrons transmit a lot of momentum and energy to the protons.

SLAC Starts Its Work

To test these hypotheses, the accelerator was built at the Stanford Linear Accelerator Center (SLAC) in California. The first experiments that looked inside the proton or neutron were conducted in 1969. The outcome was a total surprise, for it was found that the electrons frequently scattered at wide angles, losing a lot of energy, as though they were encountering small but very hard pieces on their voyage through the nucleon. (Remember that precisely the same thing had happened sixty-six years before when it was

The Proton is X-rayed in California

discovered that alpha particles can scatter off atoms at sharp angles.)

Of course, the first thought in everyone's mind in 1969 was to identify the little pieces inside the proton as the quarks. If this is the case, certain characteristics of the scattering process can be predicted. After all, the experimenters already knew the electric charge of the quarks. The point is this: the electric charge of the quarks determines the probability that an electron will scatter at a certain angle if we treat the quarks as structureless objects. A simple analysis of the experimental data confirmed this suggestion; the charges for the pieces inside the proton turned out to be 2/3 and −1/3, the same charges we found previously (figure 6.2).

Figure 6.2
High-energy electrons penetrate deep into the proton and scatter off the quarks, which act as pointlike objects. The interaction of electrons with quarks is determined by the electric charge of the quark.

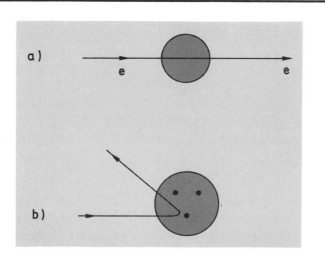

The SLAC experiments provided detailed information about the quark structure of the proton. Let us suppose that a proton consists of three loosely bound quarks. If that is the case, we would expect a proton moving at high speed to look like three quarks moving at the same speed. Specifically, we would expect each quark to carry approximately one-third of the total momentum of the proton.

Physicists conducting the SLAC experiments were able to determine not only the charge of a quark inside a proton but also the contribution the quark makes to the total momentum of the proton moving at almost the speed of light. (An electron moving through the SLAC accelerator and hitting a proton at rest sees the proton as moving at a speed close to the speed of light.)

The fractional contribution of each quark to total proton momentum is denoted by what we call the x parameter. This number has some value between 0 and 1. If the quarks inside a proton were loosely bound, the x parameter of each quark would be close to 1/3. Interestingly enough, the SLAC experiments do not support this supposition. Instead the x parameter varies from one scattering event to another. In other words, the quarks do not each carry a fixed fraction of the total proton momentum.

A fundamental law of physics states that the total momentum of a system is equal to the sum of the momenta of all its parts. Therefore, the sum of all quark momenta should equal the total proton momentum. Testing this law in the case of quarks, we come upon a surprising result. The sum of the quark momenta in any proton turns out to be significantly smaller than the total proton momentum. Almost 50 percent of the proton momentum is not accounted for by the quarks. What is to be done? Was there something wrong with the experiment, or maybe with the law of conservation of momentum? We are reminded of a similar situation, that of β decay, which we discussed earlier. In that case, Pauli "rescued" the law of conservation of energy and momentum by postulating the existence of a new particle, the neutrino. Perhaps we should postulate something new here as well.

The Glue That Holds the World Together

One way of explaining why quarks do not carry the same momentum as the proton moving at nearly the speed of light would be to postulate the existence of other objects besides quarks. If there are such "particles," they must have remained invisible in the SLAC experiments (otherwise we would have seen them) and they must be electrically neutral (otherwise they would have interacted with the electrons), since only charged constituents are visible in electron-proton scattering experiments.

It turns out that introducing other constituent particles is exactly the right move. These new particles are called gluons—and appropriately so, for the gluons, which are the "glue" that holds the world together, are what bind quarks to form hadrons.

We shall see later that gluons are essential to the dynamics of the quarks inside the proton. They have no electric charge and do not interact directly with electrons. However, they do have momentum and energy. In a fast-moving proton they carry about 50

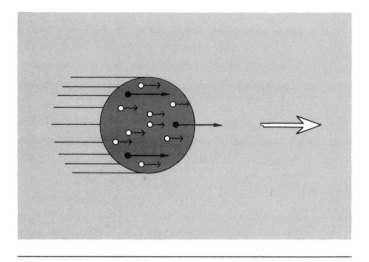

Figure 6.3
A proton moves at high speed in the direction indicated by the arrow. The total momentum of the proton is the sum of the momenta of the three quarks (black circles) and the momenta of the relatively large number of gluons (open circles). The contribution of gluons and quarks to the total momentum is about half for each.

percent of the momentum. The sum of the three quark momenta plus the momenta of the gluons account for the total momentum of the proton (figure 6.3). But just how many gluons are there inside a proton? It turns out that the number is rather large. A single gluon carries only a small fraction of the total proton momentum. It takes the contribution of many gluons to accumulate half the momentum of the proton.

The Proton is X-rayed in California

Neutrinos Test the Interior of the Nucleon

If electrons are so useful in exploring the interior of the proton, perhaps we can use neutrinos in a similar way. Neutrinos, like electrons, do not interact strongly and therefore should penetrate just as deeply inside the proton as the electrons did. The SLAC experiment with electrons explored the distribution of electric charge inside the proton. Neutrinos do not interact electrically, however, and so we can use them only to explore the proton structure with respect to the weak force. We know that the electric charge of the proton is concentrated in the three quarks, which act as centers of electromagnetic interaction. Are there also definite centers for the weak force?

To answer this question, we have to find out what happens when a neutrino hits a quark inside a proton. Although this reaction involves the weak interaction, which we have not discussed so far, we can state, somewhat prematurely, what actually happens. It turns out that often a neutrino in such a weak interaction is transformed into a charged lepton. For example, an electron-neutrino turns into an electron and a muon-neutrino into a muon. Simultaneous with these transformations, the quarks change, too. In 95 percent of all cases the d quark turns into a u quark (something else occurs in the remaining 5 percent, but this shall not concern us here right now).

The conversion of a neutrino into a charged lepton is extremely useful from the viewpoint of the experimental physicist. We have no problems observing a charged lepton, and therefore we can "see" the neutrino scattering process in the laboratory.

Figure 6.4

The Gargamelle bubble chamber at CERN. The chamber is filled
with a special liquid, for example, liquid hydrogen. The pressure
is suddenly reduced whenever a beam of neutrinos passes through
the chamber, and the interaction of a neutrino with a nucleon
inside the chamber produces a number of charged particles, with
little bubbles forming along their trajectories. The trajectories of
the charged particles can be seen in the form of tracks consisting
of strings of bubbles. Outside the chamber we can see a magnet,
which generates a strong magnetic field inside. This field bends
the trajectories of the charged particles. Information about the
charge and momentum of the particle is obtained by measuring
the direction and the degree of the bending of the trajectories.

The Proton is X-rayed in California

Special neutrino-beam facilities were constructed at several research centers, especially at CERN and Fermilab, to study the interaction of high-energy neutrinos with protons and neutrons. To see what happens during such an interaction, CERN built a large bubble chamber, called Gargamelle (figure 6.4). Like any bubble chamber, Gargamelle is a tank containing a liquid (liquid hydrogen, for example) heated to just below the boiling point and maintained at a certain pressure. Such a liquid forms gas bubbles when the pressure is reduced slightly, and the bubbles tend to form especially in regions through which an electrically charged particle has passed. Therefore, particles to be studied are injected into the chamber when the pressure is slightly reduced, creating particle trajectories in the form of a string of bubbles. (These tracks are similar to the vapor trails that airplanes sometimes leave in the sky.) A strong magnet outside the bubble chamber produces a magnetic field inside, which bends the trajectories of the particles. The particle tracks are then photographed and the curve of the trajectories measured to estimate the charge and momentum of the particle.

Results obtained around 1973 from the Gargamelle bubble chamber experiments made a useful contribution to our understanding of the interaction of neutrinos with quarks. Physicists observed that the neutrinos also interact with the quarks of the proton, just as the electrons in the SLAC experiments did. It seems that quarks are centers not only of electromagnetic interaction inside the proton but also of the weak interaction. Thus we have another tool to investigate the structure of hadronic matter (figure 6.5).

As seen in the electron and neutrino experiments,

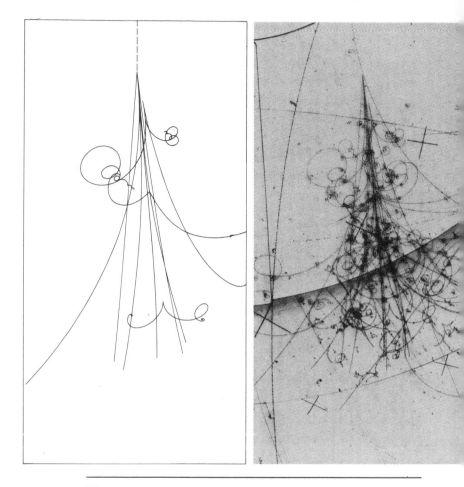

Figure 6.5
A typical event at the large European BEBC bubble chamber at CERN. A neutrino enters the chamber from above and hits a proton. The neutrino, being a muon-neutrino, turns into a muon, which moves in a downward direction. The proton is destroyed by the reaction with the neutrino, and large numbers of particles are created. On the right is the bubble chamber picture, on the left the relevant tracks.

quarks, like leptons, appear to be structureless. This is particularly surprising in view of the fact that quarks are strongly interacting particles. Somehow or other, the quarks' ability to interact strongly appears to be lost or neutralized when they encounter a high-energy electron or neutrino. Does the strong interaction disappear entirely in such an event, or is it merely attenuated? The experimental data of the early 1970s were not good enough to tell us whether quarks lacked structure entirely or whether some of the strong force remained. Nevertheless, the results obtained from the SLAC and neutrino experiments impose an important constraint on a theory of strong interactions: the theory must allow the strong force to weaken at high energies (or, correspondingly, at very small distances).

Theoretical physicists have invented a special name for a theory of this type: they call it asymptotically free, which means that at high energies quarks act almost like free particles. We shall see later that the constraint of asymptotic freedom points toward quantum chromodynamics as the correct theory of the strong interaction. First, though, chapter 7 introduces another important aspect of the quark theory: the existence of other quarks besides the u and d varieties.

VII

A Strange
New Quark

At the beginning of the 1950s, the world of particle physics was relatively simple. All that existed were hadrons and leptons, and the hadrons could all be explained in terms of two types (flavors) of quarks. If one wanted to, one could regard leptons and quarks as one unit:

$$\begin{pmatrix} u & \vdots & \nu_e \\ d & \vdots & e^- \end{pmatrix}$$

This scheme allows no room for the muon, however, and the question arises of what to do with it. The muon is a foreign body in the world of particles. The American nuclear physicist Isidor Rabi justifiably asked, "The muon—who ordered that?"

Today we know that the cosmic ray physicists who in the 1950s discovered a whole new set of particles—

particles that looked rather peculiar and were subsequently named strange particles—also found a partial answer to Rabi's question. Those discoveries opened up one of the most fascinating chapters in the annals of elementary particle physics.

The first particle that was found was the neutral Λ particle with a mass of 1116 MeV, roughly 20 percent heavier than the proton. This new particle has the strange quality of being produced copiously in hadronic collisions but decaying relatively slowly. Most of the massive particles produced in hadronic collisions (for example, the Δ particles) decay as soon as they are created (their estimated lifetime is on the order of 10^{-24}s, the time it takes a ray of light to travel through a nucleon. The Λ particle however, lives much longer, about 10^{-10}s. Moreover, this particle is always produced in the company of another particle, the K meson, whose mass is 495 MeV. This phenomenon of producing new particles in pairs is called associated production.

Strangeness—A New Quantum Number

Many different ideas were proposed to explain the existence of these new strange particles. The correct explanation is that there is another degree of freedom besides isospin in strong interaction physics. The new particles like the Λ hyperon are endowed with this quality, which is best described by a new quantum number called the strangeness number. Strangeness is a quantum number similar to the baryon number and can be ascribed to every particle. The strangeness number for normal particles, such as nucleons and pions, is zero, and the Λ particles have been assigned a strangeness value of −1. Leptons do not carry strangeness, and neither does the photon.

The hypothesis was that, like isospin, strangeness is conserved in the strong interaction. This means that during a strong interaction the total strangeness of a system does not change, and this explains why strange particles can be produced only in pairs, via associated production. The reasoning goes as follows. Since nucleons and mesons have strangeness zero, the final system formed when we scatter two nucleons or a meson and a nucleon also has a vanishing strangeness. If we produce a Λ particle in such a process, however, the strangeness of the final system would be −1, that is, strangeness would not be conserved, unless another strange particle, one with strangeness +1, is also produced. In this case, the strangeness for the system as a whole is again zero. This, in fact, is precisely what happens. If a Λ particle is produced, we can be sure of the production of another particle, generally a K meson with strangeness +1.

There is an important difference between the strangeness quantum number and the baryon number: strangeness cannot be conserved in all processes. If it were, the Λ particle could not decay, but would be stable like the proton. Yet it does decay within the relatively short time span of 10^{-10}s, into a proton and a π^- particle or into a neutron and a π^0 particle.

Investigating the decay of the Λ particle in greater detail, one finds a surprise: the decays of the Λ particles and of all strange particles are similar to the β decay of the neutron. The decays of these particles are caused by the weak interactions as well.

A Strange New Quark

We conclude that the strangeness quantum number is not conserved in the weak interaction and that in this respect strangeness is similar to isospin. Isospin is violated by the weak interaction, too, since the isospin changes during β decay (a neutron turns into a proton).

As mentioned above, strangeness, like isospin, is conserved in the strong interaction. This explains why strange particles can be produced so easily in pairs in a strong interaction but take relatively long to decay. In a strong interaction, the strange particles are created in pairs out of pure energy—one particle carrying the strangeness $+1$, the other the strangeness -1. Nothing prohibits such a process, and it proceeds rather easily. However, in the decay of strange particles the strangeness of each particle actually disappears. This process is slow because it proceeds only via the weak interaction.

Today we know that the discovery of the strange particles of the 1950s was the "premature" discovery of a new quark flavor, the strange quark, denoted by the letter s, which joins the u and d quarks. Specifically, the Λ hyperon is a system that consists of the union of three different quarks (uds). Since the electric charge of the Λ particle is zero, we can determine the electric charge of the s quark: it is $-1/3$, like that of the d quark. We interpret the s quark as some kind of "heavy" brother of the d quark, which simply means that particles containing an s quark are heavier than corresponding particles without strangeness, generally by 150 to 200 MeV.

It is occasionally helpful to use effective mass when discussing quarks. Since a proton consists of three quarks, the effective mass of the u or d quark is simply the mass of the proton divided by three: $m_u = m_d$ $= M_p/3 \approx 300$ MeV. Of course, effective mass has nothing to do with the "real" mass of the quarks, since quarks either do not exist or, if they do exist, are very heavy. If we assign the u and d quark an effective mass of 300 MeV, the effective mass of the s quark is about 450 MeV.

Strange Relatives of the Nucleons

Taking it for granted that the s quark exists along with the u and d quarks, we now ask how many and what kinds of new particles we can construct with these three flavors. To answer this question, we proceed as follows. We start with the particles made up of u and d quarks: the nucleons, the Δ particles, the pions, and the ρ mesons. We replace one of the u or d quarks with an s quark, then we replace two u or d quarks with an s quark, and so on.

First of all, let us consider what strange partners the nucleons might have. We have already mentioned one of them, the Λ hyperon, but we expect more, specifically, one particle corresponding to a uus quark combination and having charge +1 and one particle corresponding to a dds combination and having charge −1. As is immediately apparent, these two particles are related via the exchange of u and d. We obtain dds from uus by exchanging the u with the d. Since the u-d exchange is an isospin transformation, the two particles must belong to a group of particles that are related to each other via isospin symmetry, in other words, to an isospin multiplet.

A Strange New Quark

Recall that we have already made the acquaintance of an isospin multiplet containing a particle with charge +1 and another with charge −1, namely, the isospin triplet π^+, π^-, π^0 (the pion multiplet). Therefore we expect our new particles to belong to an isospin triplet as well. The third particle, then, must have charge zero and must also contain an s quark. There is only one possibility—the third and neutral partner must have the composition uds, that is, the same composition as the Λ particle. In principle, the neutral particle could be identical to the Λ hyperon. However, this is not the case. There is a second particle with the quark composition uds. This particle is called Σ^0, and its charged isospin partners are Σ^+ (uus) and Σ^- (dds). Σ hyperons were discovered at the beginning of the 1950s in cosmic rays. They are heavier than the Λ hyperon, having a mass of about 1190 MeV.

This analogy between pions and Σ hyperons refers only to their isospin content; in all other respects these particles are completely different. One essential difference is that the antiparticle of the π^+ is the π^-, and the π^0 is its own antiparticle, while the antiparticles of the Σ hyperons are three new particles, denoted by $\overline{\Sigma}^-, \overline{\Sigma}^0, \overline{\Sigma}^+$. Thus the Σ hyperons plus their antiparticles are a system consisting of six particles.

The Λ particle can be interpreted as a kind of modified neutron in which one of the d quarks has been replaced by an s quark. It does not have isospin partners, but instead is a lonely particle, an isospin singlet. What happens if we remove both d quarks from the neutron and replace them with s quarks, ending up with a uss quark configuration, which is a particle of strangeness −2? Such particles do indeed exist and have been named Ξ hyperons. Obviously there are two such particles, namely, Ξ^0 (uss) and Ξ^- (dss), which form an isospin doublet. The mass of the Ξ particle is about 1320 MeV. This is larger than the mass of a Σ particle, of course, because the Ξ particle contains two s quarks.

Experiments show that Ξ particles live roughly as long as either Λ or Σ particles, that is, about 10^{-10}s. This indicates that the decay of Ξ particles is also caused by the weak interaction. It turns out that the Ξ particle always disintegrates by emitting a Λ hyperon. For example, the Ξ^- can decay into a $\Lambda\pi^-$ system. By looking at Λ decay in detail, we have learned that a change in strangeness from -1 to 0 takes place during the process, that is, the strangeness changes by one unit. One might think that something similar is true for the Ξ hyperons, and this is indeed the case. It turns out that the Ξ particles have strangeness -2. If the strangeness changes in the decay process by one unit, one expects the final state to have strangeness -1. It becomes clear why the Ξ particle always decays into a particle system that contains a Λ particle. It cannot decay into a nucleon right away because nucleons have strangeness zero ($\Xi^0 \rightarrow p\pi^-$ does not exist). Of course, the Λ particles emitted in Ξ decay subsequently into nucleons. Thus the weak decay of a Ξ particle looks like a cascade. First we produce a system of strangeness -1, and later we arrive at particles of strangeness zero (figure 7.1). In each step, the strangeness changes by one unit.

If we consider the strangeness of these particles (-1 for Λ and Σ, -2 for Ξ) and at the same time the number of strange quarks each contains, we discover a parallel. The number of strange quarks is precisely

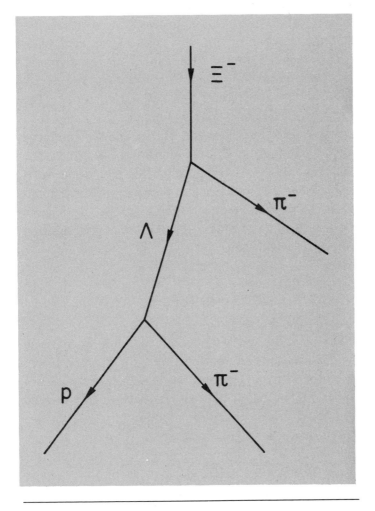

Figure 7.1
A diagram of Ξ decay. The Ξ⁻ particle decays first into a Λ and a
π⁻. (This system has strangeness −1.) The Λ particle then decays
into a proton and a π⁻. (This system has strangeness 0.) The
strangeness changes in a cascading fashion.

equal to the negative strangeness.* It is instructive that the number of strange quarks in a hyperon is directly related to the strangeness of the hyperon. The conservation of strangeness in the strong interaction can be understood in very simple terms. The number of strange quarks remains constant in a strong interaction. Strange quarks can be produced only in pairs, as an s quark and an anti-s quark. The pair production of strange particles in a collision is nothing but the production of a s$\bar{s}$ pair. After the collision the s quark resides inside a strange particle—a Λ particle, for example—and the $\bar{s}$ quark is emitted as part of a K meson.

The Eightfold Way to Form a Baryon

Let us count the number of particles we have so far. Besides the two nucleons, we have the Λ hyperon, the three Σ hyperons, and the two Ξ particles. Altogether we have eight baryons. Is there by any chance something special about the number 8?

Ten years after the discovery of the first strange particle, Murray Gell-Mann and Yuval Ne'eman realized that a very useful description of the particle system in hadronic physics can be derived by generalizing the mathematical methods describing the isospin properties of the particles. Such methods, widely used in physics, are called group theory. The mathe-

---

*This amounts to a somewhat unfortunate accident in physics. It would be preferable to have defined the strangeness of these particles the other way around and to have given the Λ particle a strangeness number of +1. However, at the time the strange particles were discovered, no one knew about quarks. So it happened that the "wrong" sign was chosen for strangeness.

matics of group theory, at least those parts relevant for our purpose, was developed largely during the last century by the French mathematician Elie Joseph Cartan.

Gell-Mann and Ne'eman used Cartan's method to extend isospin to a higher symmetry until they arrived at what is called unitary group SU(3). In today's language, SU(3) is nothing but a shorthand notation for describing the various ways of combining quarks to form hadrons. Using the methods of Gell-Mann and Ne'eman, one can predict the existence of eight baryons, the very same eight particles mentioned above. The unit comprising those eight baryons is called the baryon octet (figure 7.2).

Figure 7.2
The eight baryons consisting of u, d, and s quarks. They form an octet under the symmetry group SU(3).

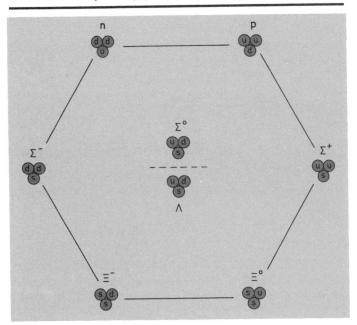

What about the mesons? With three quarks and three antiquarks we can construct $3 \times 3 = 9$ possible $\bar{q}q$ configurations. Specifically, we expect to find particles of strangeness 1 or -1 that correspond to the quark configurations $\bar{u}s$ or $\bar{s}u$, and so forth. These are the previously mentioned K mesons, which are produced in association with hyperons in hadronic collisions. Using the language of SU(3) group theory, spin-zero mesons like the π mesons form an octet plus a singlet, that is, nine particles altogether (figure 7.3)

It is a special feature of the strong interaction between quarks that the neutral mesons lack a unique

Figure 7.3

The nine mesons. The neutral mesons of strangeness 0—π^0, η, η'— are mixtures of the quark configurations $\bar{u}u$, $\bar{d}d$, and $\bar{s}s$. This is denoted by the circle around these configurations.

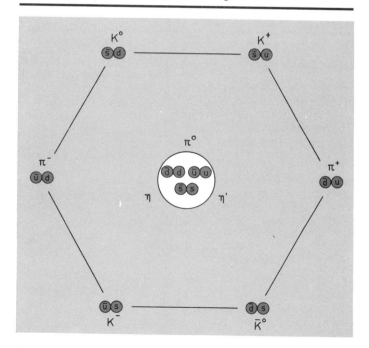

quark composition. For example, the neutral pion is a mixture of ūu and d̄d: 50 percent of the time the π^0 is a ūu system, and 50 percent of the time it is a d̄d system. There is nothing wrong with this. In quantum mechanics we often deal with such mixtures.

Of course, strange partners of ρ mesons exist, too. These are vector mesons (mesons with spin 1) with the quark composition ūs and d̄s. They are called K*. Their masses are 892 MeV. In addition there is a spin -1 particle with the quark composition s̄s, that is, it is composed of a strange quark and its antiquark. This particle is called phi (hereafter denoted by Φ); its mass is 1020 MeV. Like the spin-zero mesons, the vector mesons also form an SU(3) octet as well as a singlet, that is, there are nine vector mesons in all.

The Hunt for Omega Minus

Thus far we have discussed only a small portion of all the strange particles that have been discovered. We mention only one more set of particles, the strange partners of the Δ baryons (figure 7.4). In all, there are ten baryons of spin 3/2, which form a decuplet. One of these particles is very strange indeed—it consists of three strange quarks. It has, therefore, strangeness -3 and electric charge -1 and is called Ω^-.

Not all particles needed to fill the baryon octet and decuplet were known when the SU(3) theory of strongly interacting particles was developed. Specifically the Ω^- particle had not been discovered. The theorists were able to predict not only the existence of

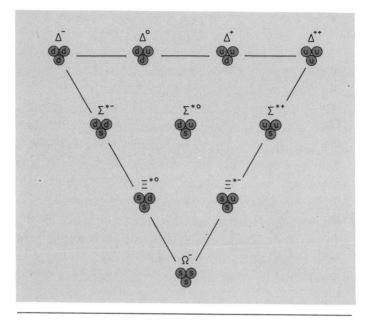

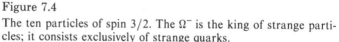

Figure 7.4
The ten particles of spin 3/2. The Ω^- is the king of strange parti-cles; it consists exclusively of strange quarks.

the Ω^-, but also its mass, which they thought would be 1670 MeV, almost twice that of the proton.

Finding the Ω^- particle in an experiment clearly was of crucial importance for the validity of both SU(3) theory and quark theory. The hunt for the Ω^- started at the beginning of the 1960s. In 1964 Ω^- was found by a group at Brookhaven National Laboratory in New York. It had a mass of 1672 MeV, and its discovery was one of the great breakthroughs in the development of particle physics. It confirmed to both theorists and experimenters that they were on the right track.

A Strange New Quark

Quarks and Leptons

Let us summarize the situation in particle physics as it was in 1964. All hadrons, be they baryons or mesons, discovered by that time were explicable as bound states involving the three quarks u, d, and s. One may suppose that there exists a close parallel between leptons and quarks. In fact, the leptons appear analogous to the quarks: we have two light leptons, e^- and ν_e, which correspond to the light quarks u and d, and a heavier lepton, the muon, which corresponds to the heavier quark.

However, this parallel between leptons and quarks was not as obvious for the following reasons. By the end of the 1950s many theorists suspected the existence of another neutrino in addition to the electron-neutrino, specifically, a neutrino related to the muon by the weak interaction, just as the electron is related to its neutrino. Direct experimental proof of the existence of this other neutrino, the muon-neutrino, came a few years later, in 1961, from Brookhaven. From that point on, physicists had to think in terms of four leptons, not three. Thus three quarks were lined up against four leptons:

$$\begin{pmatrix} u & \vdots & \\ d & \vdots & s \end{pmatrix} \qquad \begin{pmatrix} \nu_e & \vdots & \nu_\mu \\ e^- & \vdots & \mu^- \end{pmatrix}$$

This certainly was not a very appealing picture. Asymmetric to say the least. What is missing here? Perhaps there exists yet another quark? Theorists had even named the still missing fourth quark: charm.

But does charm exist? This question leads us directly to a discovery made in 1974. We are entering the era of contemporary particle physics.

VIII

Particles with Charm and a New Force

November 11, 1974, at Cal Tech was just another Monday for me until my colleague Richard Feynman surprised me with the news that, over the weekend, workers at SLAC had discovered a new particle with highly unusual properties. This new particle lived about 10 000 times longer than a normal particle that decays in the strong interaction (recall that the typical lifetime of a particle that can decay via the strong interaction, the Δ particle, for example, is on the order of 10^{-24}s) and had a mass of about 3100 MeV. This particle had a quality never before encountered in physics—but what was it?

Like most physicists in the United States and Europe, we held informal meetings at Cal Tech that memorable day and discussed the most likely interpretation of the new phenomenon, which was called charm.

The Weak Decay of Mesons

To understand why charm is a useful property, we must once again consider the weak interaction. When a high-energy neutrino reacts with matter, it normally produces a charged lepton, either an electron if the neutrino is an electron-neutrino or a muon if the neutrino is a muon-neutrino (figure 8.1). A similar process is responsible for the weak decay of hadrons. The

Figure 8.1
A normal neutrino reaction. A muon-neutrino reacts with a proton and becomes a muon. A large part of the energy of the incoming muon-neutrino is absorbed by the proton, which consequently "heats up" and disintegrates into several particles.

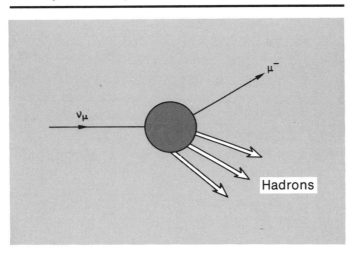

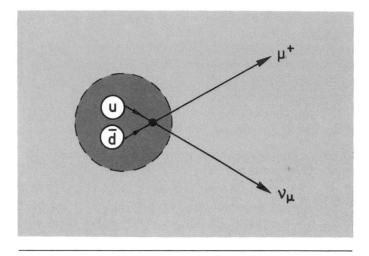

Figure 8.2
The decay of the positively charged π^- meson. The u and $\bar{\text{d}}$ quark annihilation produces a neutrino and a positively charged muon.

decay of a π^+ pion, for example, is accounted for by the annihilation of the quark-antiquark pair inside the pion to produce a muon and a muon-neutrino (figure 8.2). The decay of a K meson into a muon is described in analogous fashion, and the same force is responsible for the decay of K^+ and π^+. These decays have only one slight difference: the K decay proceeds about twenty times more slowly than expected naively. We interpret this by saying that the main part of the weak force acts on the u and d quarks, whereas s quarks participate less intensively in the weak force. The reason for this difference is to be sought in the structure of the weak interactions themselves.

At the beginning of the 1960s, some physicists were already asking whether perhaps there was *another* weak force, one that allowed a neutrino to react with matter without the neutrino itself being transformed

Particles with Charm and a New Force

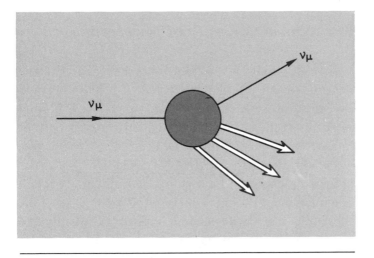

Figure 8.3
A neutrino reaction in which the muon-neutrino does not change its identity, called a neutral current process. The incoming muon-neutrino transmits a large portion of its energy and momentum to the hadron (for example, a proton inside an atomic nucleus). The hadron is thus excited and disintegrates into several particles. The number of the particles produced in this hadron decay depends on the energy transmitted.

into an electron or a muon. Such a weak force would be like the electric force. An electron reacting with matter, with an atomic nucleus, for example, via the electromagnetic interaction does not change its identity. It may change its energy and momentum, but it remains an electron. Therefore, one might also expect a neutrino to react with matter without changing its identity (figure 8.3) in a process called, for reasons that will become clear later, the neutral current interaction.

The Neutral Current Is Discovered

In 1970, only a few physicists believed that the neutral current interaction existed. If it did, we would expect to observe it just as we can observe normal weak interactions (sometimes called charged current interaction), namely, in the weak decays. Specifically, we would expect the neutral K meson, which consists of an s antiquark and a d quark, to disintegrate into a pair of muons via the new type of weak force. Furthermore, this decay should happen at roughly the same rate as the decay $K^+ \rightarrow \mu^+ \nu_\mu$. Experiments searched for such decays, but to no avail. The data are very precise. Today we know that only 1 out of 100 million K^0 decays proceeds through the emission of a pair of muons. Therefore the conclusion was reached that the new speculative neutral force does not exist.

Actually this argument is not airtight. The experimental data for the decay of the neutral K mesons merely imply that the neutral force never affects a d and an s̄ quark at the same time. In other words: no problem would exist if the neutral force never changed a quark flavor, for example, never changed d into s or vice versa. A neutral force that does not do anything of this kind is permitted.

In fact, it is quite natural to think that the neutral current interaction will not change quark flavor. After all, the same is true for the electromagnetic interaction, where a photon reacts with leptons or quarks without changing one type of lepton into another (for example, an electron into a muon) or a d quark into an s quark. If the neutral current interaction has the same property, there is only one possibility of ascertaining it, and that is through the study of neutrino processes at high energies.

The search for evidence of the neutral current interaction was one of the main tasks of the Gargamelle collaboration. In 1973, after more than a year of experimenting with the Gargamelle bubble chamber at CERN, the situation was finally clarified: neutrinos sometimes react with hadrons without changing into charged leptons. That the neutral current interaction did indeed exist was beyond doubt. Shortly afterward, FNAL also found indisputable evidence for the existence of the neutral current interaction. Its properties will be discussed in greater detail later on.

Now that its existence had been established, the question arose, What mechanism prevents the neutral current force from contributing to the decay of the neutral K meson? Only one mechanism of this kind was known at the time of the discovery of the neutral current force, and this mechanism required the existence of a new quark flavor of electric charge 2/3.

The odd asymmetry between leptons and quarks troubled theoreticians as early as 1964, as mentioned in Chapter 7. If indeed there was some anology between quarks and leptons, then three quarks versus four leptons did not seem to make much sense. The simplest way of creating a better anology was to add a new quark with charge 2/3. This would be the charmed quark or c quark, some kind of heavy brother of the old u quark. Introducing the c quark would allow us to construct the following scheme:

$$\begin{pmatrix} u & | & c \\ d & | & s \end{pmatrix} \qquad \begin{pmatrix} \nu_e & | & \nu_\mu \\ e^- & | & \mu^- \end{pmatrix}$$

However, at this stage the addition of charm was simply one of many ways of enlarging the system of leptons and quarks. For most physicists, it was just one way of arriving at lepton-quark symmetry—simple but by no means convincing. Not until 1970, six years later, was it realized that charm was also useful in ensuring the stability of the neutral K meson so that it would not be affected by decay through the neutral weak force. The argument, as presented by Sheldon Glashow and his collaborators at Harvard, is extremely simple: if charmed quarks really exist, then the normal weak interaction (that which involved a change of the electric charge and is responsible for the decay of the neutron) must be such that it acts on the ud quark system in the same way as it acts on the cs quark system. In this case, the interaction leading to the decay of the neutral K meson into leptons receives a contribution both from the weak force acting on the ud system and from the weak force acting on the cs system. The two contributions are of exactly the same magnitude but opposite in sign. As a result, the neutral weak force never changes one type of quark into another and the neutral K mesons cannot decay into leptons.

If the neutral current interaction actually exists, then it becomes essential that charm exists, too. But at what level of energy and where might it appear? It was clear that the effective mass of the c quark had to be much greater than that of the s quark; otherwise, the former would already have made its appearance.

The question became, What new particles would we expect to exist in the presence of a charmed quark? First of all, we expect a new degree of freedom to manifest itself in a somewhat frozen state, that is, in the form of a $\bar{c}c$ state—the new quark and its own antiquark. In analogy to positronium, which is the bound state of an electron and a positron, such states have been named charmonium. Such "onium" states are familiar to us for s quarks; for example, the Φ meson is an $s\bar{s}$ state. Furthermore, we expect the existence of bound states (mesons) consisting of a charmed quark and one of the old u and d quarks. These particles would carry the new charm quantum number (the number of c quarks minus the number of $\bar{c}$ quarks), and they would be stable as far as the strong interaction is concerned. For example, the new mesons called D mesons, like the K particles, can decay only via the weak interaction. This implies that the charmed mesons live for a relatively long time—much longer than, for instance the $\bar{c}c$ mesons, which can decay via the strong interaction. Figure 8.4 shows

Figure 8.4
The new mesons predicted by the charm scheme.

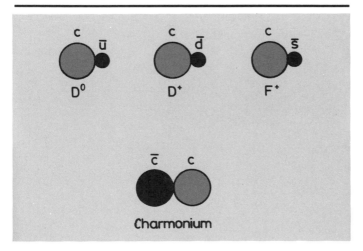

the various mesons that contain c quarks.

Of course, the charmed quark can also bind to an s̄ quark, in which case one obtains a meson with electric charge +1. These mesons are called F mesons (figure 8.4). Furthermore, the c quarks can bind to two other quarks to form a baryon (for example, cud) carrying the new charm quantum number. In any case, it is apparent that the charm theory has certain prophetic qualities. It predicts the existence of a large number of new particles with well-defined physical properties. Once all this was realized by theoreticians and experimentalists in 1973, the main question became, At what energies will the new particles manifest themselves? The search for charm was on.

On the Road to Discovery

Around 1970 a group led by Leon Lederman conducted an interesting experiment at Brookhaven National Laboratory. The group directed an intense proton beam on a target and then went on the lookout. Specifically, they kept watch for the production of $\mu^+\mu^-$ pairs. Because vector mesons sometimes decay via electromagnetic interaction into $\mu^+\mu^-$ pairs, the Lederman group could study the production of vector mesons in hadronic collisions rather precisely, which could not be done easily any other way. They also hoped to find new types of particles by looking at the $\mu^+\mu^-$ pairs.

The point is this. A new particle that decays into a $\mu^+\mu^-$ pair produces the pair in such a way that the mass of the pairs corresponds precisely to that of the original mass (the mass of a pair of particles is defined as the energy of the pair in the rest system). Therefore it was possible for the first time to conduct a careful study of the mass spectrum of the dimuon pairs up to energies of more than 4000 MeV. No new particles were found, but a strange kink appeared in the dimuon mass diagram at about 3000 MeV. Something odd seemed to be going on at this energy.

Lederman himself felt that the effect was due to a new particle and that he simply could not see it because of the uncertainty of his measurements of muon momenta. So he decided to build a new particle detector to investigate the matter further at a later time using the new accelerator at Fermi Laboratory.

Meanwhile, another group at Brookhaven, this one under the leadership of Sam Ting (figure 8.5), was conducting a similar experiment. Instead of looking for muons, however, the Ting group, using a very so-

Figure 8.5

Professor Samuel Ting, currently at the Massachusetts Institute of Technology. He was the leader of the Brookhaven National Laboratory group which discovered the J/Ψ particle in proton-nucleon collisions in 1974.

phisticated apparatus, was concentrating on the detection and identification of electrons. In autumn 1974 they began their search for electron-positron pairs produced in proton-nucleon collisions in the mass region between 2000 and 4000 MeV. They found, to their surprise, that many electron-positron pairs have a mass of about 3100 MeV. After a few weeks of analysis, it became clear that there had to be a particle of mass 3100 MeV decaying into electron-positron pairs.

At the same time, another experiment was being conducted in California, where the electron-positron ring called SPEAR (figure 8.6) had been completed. This machine, which is part of SLAC and is primarily an aggregate of appropriate magnets, consists of a positron beam and an electron beam running in opposite directions. These beams collide with each other at certain intervals and produce new particles.

In a certain sense, the experiments at SPEAR were the opposite of those Ting was conducting. If a particle is produced in a collision of protons and decays into an electron-positron pair, the same particle can also be produced via electron-positron annihilation. One of the advantages of the annihilation experiments is that the energy of the electrons and positrons can be "fine-tuned" to produce a particle of a particular mass, without any bothersome side reactions. To find such a particle, however, one has to scan the particular energy region very carefully, and this is exactly what was done at SPEAR in November 1974, especially for the energy region around 3100 MeV. Without knowing that Ting's group in Brookhaven had recently observed a particle of mass 3100 MeV,

Figure 8.6
The storage ring SPEAR at the Stanford Linear Accelerator
Center (courtesy SLAC).

the physicists at SPEAR gradually worked their way
up to this energy level, finally reaching an energy of
3097 MeV the second weekend in November 1974.
Suddenly the particle detector started to go wild—the

the small motions required by the laws of quantum mechanics). The spins of the two quarks are aligned, and so the total angular momentum of the state is +1. This is essential, since otherwise the J/Ψ state could not have been discovered in the e⁺e⁻ annihilation experiments; only mesons with spin 1 can disintegrate electromagnetically into electron-positron pairs.

But the spin of the c and c̄ quarks can also be opposite each other, in which case one obtains a pseudoscalar particle. Theorists predicted such a state (called η_c) shortly after the discovery of J/Ψ. It was expected that the mass of this η_c state would be slightly less than that of J/Ψ and that it would become visible during the disintegration of J/Ψ into a photon and the new state (figure 8.9). In 1979 the η_c particle was found this way at SLAC. Its mass turned out to be about 2980 MeV.

Although the existence of the spectrum of the char-

Figure 8.9

The J/Ψ and η_c. The J/Ψ state is the charmonium state, in which both quark spins point in the same direction and thus the total angular momentum of the state is 1. In η_c, the quark spins are opposite each other and thus the total angular momentum is 0.

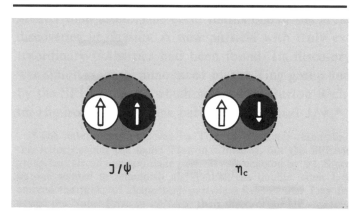

J/ψ η_c

monium states as displayed in figure 8.8 is a clear indication that the new particles consist of a new quark and its antiquark, final proof of the charm idea had to await confirmation through the discovery of particles consisting of the new quark and one of the old quarks. Theorists anticipated that the lightest of these new particles would be a neutral particle consisting of a c quark and a ū quark. This particle was called the D^0, and it has the quark composition ūc. These new mesons carry the charm quantum number, and therefore they cannot decay like the J/Ψ particle via the strong or electromagnetic interaction. Instead they can decay only via the weak interaction. Theorists estimated that these new charm particles live a hundred million times longer than J/Ψ.

At the beginning of 1975, all high-energy laboratories around the world began the search for the new charmed particles. But where was one to look for them? Which experiment and which laboratory was best suited for this task? These were open questions in 1975, and yet physicists were convinced that it was most likely that charmed particles would be created in the e$^+$e$^-$ annihilation, the same process that had best revealed the J/Ψ meson.

What can we say about the decay of charmed particles through the weak interaction? The c quark was introduced as the partner of the s quark in the weak interaction, and therefore we expect the c quark to decay by transforming into a s quark. This implies that the D meson decays into a system of particles whose strangeness is nonzero, that is, a system containing a K meson. A possible particle system would be Kπ or Kππ. Sometimes the D meson should also decay into a K meson and a lepton pair, for example,

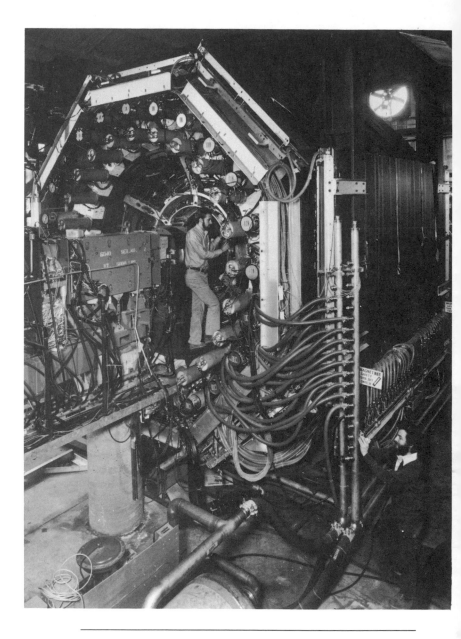

Figure 8.10
A look into the most successful particle detector in the history of particle physics, the Mark I detector at SLAC. Using this apparatus, SLAC physicists discovered the charmonium states and the D mesons (courtesy SLAC).

$Ke^-\bar{\nu}_e$. This decay affords us the opportunity to look for charmed particles by searching for the production of electrons or muons in various reactions.

In 1976, after almost two years of intensive research, the SLAC physicists succeeded in identifying the charmed particles (figure 8.10) by studying the $K\pi\pi$ particle systems produced in the e^+e^- annihilation at high energies. The masses of these systems,

---

Figure 8.11

An event at CERN in which a D meson was produced in a neutrino-hadron collision. On the left is a photograph from the BEBC bubble chamber where the event occurred; on the right, the relevant tracks. The (invisible) incoming neutrino reacts with one of the quarks inside a proton in the bubble chamber and turns into a muon (visible). The proton is excited and decays into a D meson and a proton. The D meson decays into $K^-\pi^+\pi^+$. All these particles are identifiable inside the bubble chamber. (The numbers denote the momenta of the particles in GeV.)

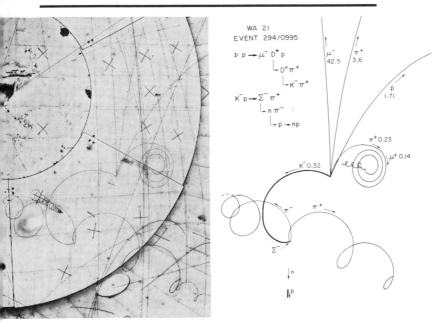

that is, the energies in the corresponding rest systems, peak at 1.86 GeV. Special experiments to investigate this in detail confirmed the original hunch that these states are the decay products of the D mesons. Shortly afterward, physicists at DESY in Hamburg found further evidence of the existence of the charmed particles when they observed the decay of D mesons into a K meson and leptons.

Subsequently, evidence has been found for the production of charmed particles in many other reactions, for example, in proton-proton collisions and in neutrino reactions (figure 8.11). Moreover, evidence has also been found for the existence of the s̄c meson (called the F meson). Table 8.1 summarizes the main properties of the D and F mesons.

TABLE 8.1
The D and F Mesons

Particle	Mass	Decay Products
D°	1863 MeV	Mostly K mesons plus other particles (Kπ or Kππ)
D⁻	1868	Same as D°
F	≈ 2040	Pairs of K mesons plus π mesons, η mesons plus π mesons, and others.

Sophisticated experiments to measure the life-span of the charmed mesons show that they live about 10^{-12} to 10^{-13}s; that is, their life-span is more than one hundred times shorter than that of the hyperons. These values are in accord with theoretical expectations.

But what about the charmed baryons? Earlier we

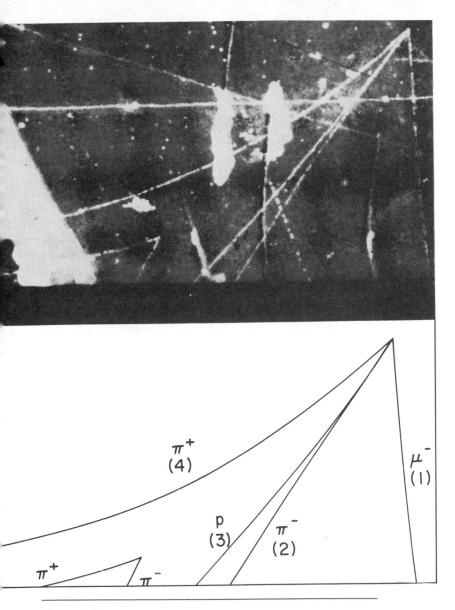

Figure 8.12
One of the events in which a charmed baryon is produced in the collision of a neutrino with nuclear matter (found at the Brookhaven National Laboratory). The incoming myon neutrino (invisible) turns into a myon, whose track is seen clearly. (The photograph shows the tracks in the bubble chamber; the relevant particle tracks are depicted schematically below.) Via the weak interaction one of the nucleons in the target is turned into a Λ^+_c particle, which decays into $p^+\pi^+ + \pi + \bar{K}^0$. The neutral K meson is not visible, and can only be seen through its decay products $\pi^+ + \pi^-$.

remarked that the lightest strange baryon is the Λ particle, which has the quark composition uds. With respect to the strong interaction, one expects the charmed quark to be simply a heavier partner of the strange quark. Let us take the Λ particle and perform a little operation by removing the strange quark from it and replacing it with a c quark. When we do this, we obtain a charmed analog of the Λ particle with the composition udc. This particle has been named Λ_c. Note that the electric charge of the Λ_c state is not zero as that of the regular Λ state is. Instead it is +1. In analogy to the situation in the strange particle system, we expect Λ_c to be the lightest charmed hadron.

Physicists searching for the charmed baryons did not find conclusive evidence for their existence until 1979. Specifically, the Λ_c particle was found to have a mass of 2273 MeV. It decays, for example, into $\Lambda\pi^+\pi^+\pi^-$ or $pK^-\pi^+u$ (figure 8.12).

The Creation of New Matter

The discovery of the Λ_c particle in 1979 marks the end of the first part of this new chapter in particle physics, which began so excitingly on a weekend in November 1974. Since 1979, we have discovered a whole spectrum of new particles that carry charmed quarks. A new type of matter has been found, one that does not exist anywhere in nature except under the extreme conditions created in high-energy physics laboratories. It is certain that these new discoveries have brought physicists closer to the innermost secrets of the universe.

IX

Red, Green, and Blue Quarks

Although quarks play a fundamental role in our understanding of hadrons and their interactions, they do not seem to exist as independent particles, as pions or protons do. What is the reason for this? Experiments so far indicate that if quarks existed as independent particles, they must be at least ten times as heavy as protons. Another peculiar feature of the quark model is that hadrons are either three-quark systems (baryons) or quark-antiquark systems (mesons). You might think that other configurations could exist, too—for instance, two-quark or four-quark systems. Of course, such particles would have nonintegral electric charges (for example, the two-quark state uu would have electric charge 4/3). Like quarks themselves, however, such states have not been found either.

Another peculiarity of quark theory can be ex-

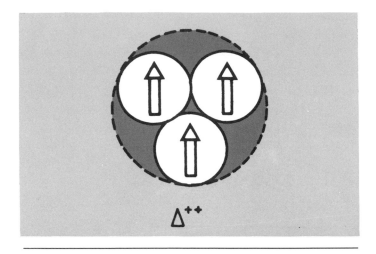

Figure 9.1
The Δ^{++} particle consists of three quarks whose spins point in the same direction.

plained best by using the Δ^{++} particle as an example. This particle has a mass of 1232 MeV and consists of three u quarks. It has 3/2 angular momentum because the spin of three u quarks all point in the same direction (figure 9.1).

The Δ^{++} particle is the lightest particle of charge 2 that exists. Therefore we expect the three u quarks to be at rest with respect to each other, aside from small motions required by the laws of quantum mechanics. (This at-rest state is called the ground state of the system.) Suppose now that we take two of the u quarks shown in (figure 9.1) and have them exchange positions. Since there is no change as a result of this switch, we conclude that the Δ^{++} configuration is symmetric.

Now, the peculiarity we have set out to explain arises as follows. Quarks are spin 1/2 particles like electrons, and spin 1/2 particles generally obey the Pauli exclusion principle, which states that two spin 1/2 objects can only be in an antisymmetric configuration. It has always been thought that all spin 1/2 objects in nature obey this principle (see our previous discussion of atomic physics). Consequently, the Δ^{++} configuration should be antisymmetric with respect to the exchange of two u quarks, and not symmetric the way it seems to be. This discrepancy represents the major paradox of quark theory, and it has puzzled physicists since the theory was proposed in 1964.

Three Colors for the Quarks

Many different ways of explaining this puzzling aspect of the Δ^{++} state have been proposed, but none of them have worked except one. The peculiarity of the Δ^{++} configuration could be related to an exception to the Pauli exclusion principle. It may be that this principle does not apply to quarks.

This is precisely what was proposed by the American physicist W. Greenberg around 1965, and shortly afterwards in a different form by J. Han and Y. Nambu. It was observed that it is possible to formulate a new type of exclusion principle (called parastatistics of rank three) that eliminates the Δ^{++} problem. Later, in 1971, this suggestion was followed up and modified by Murray Gell-Mann and myself. This modification ultimately led to the concept of quark color and to the modern theory of the strong interaction called quantum chromodynamics.

There is a simple way of circumventing the prob-

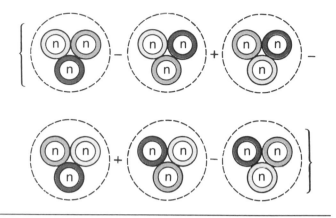

Figure 9.2
The quark configuration of the Δ^{++} particle consists of six different terms with alternating signs. The sum of all terms is antisymmetric with respect to the interchange of two quarks. (The three
colors are represented by three different shades of gray.)

lem encountered with the Pauli exclusion principle,
and that is to assume that each quark comes in three
different "editions," which we can call colors. Thus
we say that there are red and green and blue u
quarks.

Now we can construct the Δ^{++} configuration using
u quarks of three different colors, as shown in figure
9.2. How does this solve the problem encountered
with the Pauli exclusion principle? If each quark has
three colors, it is easy to arrange things so that the
Δ^{++} configuration is antisymmetric with respect to the
exchange of quarks. We simply have to make sure
that this configuration is antisymmetric in the color
quantum number. To do this, we write the Δ^{++} configuration as a superposition of six different terms (figure 9.2). Each of these terms is related to the following one by the switched positions of two colored

quarks. For example, the second term is obtained from the first by exchanging the red and green quarks and leaving the blue quark in place. Since the signs of the six terms alternate, the sum of all terms is antisymmetric with respect to the exchange of any pair of quarks, as required by the exclusion principle.

We can apply this idea in a generalized fashion by requiring all baryons, in other words, all quarks configurations, to be totally antisymmetric with respect to color. For example, the proton configuration can be obtained from the Δ^{++} configuration shown in figure 9.2 by replacing one of the u quarks with a d quark and rearranging the spins in such a way that a spin $1/2$ particle is obtained.

Baryon configurations have one property in common: each color enjoys the same rights as every other color. There is no color preference. For example, red quarks are as numerous as green quarks. Mathematicians have a special name for such configurations: color singlets. In other words, these configurations are singlets under the color group that is the group of all mathematical transformations possible in color space. Since there are three colors, this group is called SU(3), which, it turns out, is the same group that plays a role in the meson and baryon spectrum we mentioned in chapter 7. There, however, the SU(3) symmetry was used to describe all possible arrangements of the three quark flavors u, d, and s. Now we are using it to describe all possible arrangments of the three colors.

Colors and Flavors

To avoid any misunderstanding, we shall introduce the following notations. We have to distinguish between the three colors red, green, and blue and the various quark species u, d, s, c, and so forth. For quark species, we shall use the word *flavor,* just as we have been doing. Each quark can be labeled by two different indices, color and flavor. The color index always runs from one (red) to three (blue), and the flavor index denotes u, d, s, c, and any other quarks that may exist.

There is an important distinction between the color and flavor indices. The color index has only three entries. Note that our use of the concept of color makes sense only if there are exactly three colors. The problem with the Pauli principle in the Δ^{++} state cannot be solved if we have only two colors, or four, or more. The "threeness" of the color quantum number is intimately related to the fact that the baryons consist of three quarks. No similar argument can be made to restrict the number of quark flavors, however. The world of the strong interaction would be quite intact if there were only two quark flavors, and it is even possible to have only one flavor (but, of course, in that case there would be no proton). The lightest baryon in this fictitious world would be the Δ^{++} particle, which would be as stable as the proton is in the real world. As of now, we do not know how many quark flavors there actually are, just as we do not know how many leptons there are.

Since we know that baryons can be regarded as color singlet configurations, the question arises as to

Red, Green, and Blue Quarks

whether it is possible to arrange meson configurations in the same way. This turns out to be an easy question to answer—all we need to do is make sure that each color of quark occurs as often as the two others. Therefore we assume that mesons are superpositions of three different quark-antiquark terms, one composed of a red quark and its antiquark, a second composed of a green quark and its antiquark, and a third composed of a blue quark and its antiquark. We arrive at the configuration displayed in figure 9.3. We

Figure 9.3
The structure of mesons in the theory of colored quarks. A meson is the sum of three configurations: red + green + blue. The sum of all three terms is a color singlet, which can be called a "white" state.

note that all three colors again occur with equal frequency because we are again dealing with a color singlet configuration.

Using the color quantum number, we arrive at a very simple way of forming hadrons. Let us suppose that all hadrons are singlets under the color group described above. We then arrive at precisely the configuration we actually observe in nature: baryons as three-quark configurations and mesons as quark-antiquark configurations. All other configurations are excluded. Obviously, a two-quark system can never be in a color singlet state. For example, the two-quark configuration red-green does not contain blue. In other words, a two-quark configuration is not a color singlet configuration since one of the colors is always distinguished from the others. The same holds true for four-quark systems. They cannot be in a singlet state either. The only possible simple color singlets are quark-antiquark (the meson configuration) and quark-quark-quark (the baryon configuration).

Someone wanting to describe the observed mesons and baryons would say, "Let us introduce three colors and postulate that all observable hadrons are color singlets, that is, colorless ("white") objects." Asked about what he or she would do with all *other* configurations (the quarks themselves, two-quark or four-quark systems, and so forth), this person might say, "I don't know right now, but perhaps there exists an as yet unknown mechanism that prohibits real particles from having color. I notice that the hadrons I have seen are color singlets, and I take this as a hint that only color singlet states exist as physical particles. Therefore I postulate that color never appears as an isolated quality in nature; colored particles like the quarks are permanently confined together with other colored objects to form color singlets. Of course, at this moment I don't know whether this confinement postulate really holds true. I'm just trying it out as a working hypothesis. I have every intention in the world of returning to the problem later on, to see if there is a mechanism, perhaps even a simple theory, that has all the features I really want."

This was the kind of attitude some physicists took in the early 1970s to construct a sensible theory of hadrons. In physics it is quite customary to make such assumptions and to study their consequences without knowing whether the assumptions are correct. For example, Bohr assumed that electrons can move inside an atom only in very special orbits, thus defying the then current laws of classical physics. He did not know *why* electrons behaved this way but decided to disregard this question for the time being. Instead he studied the consequences of his hypothesis and found the right expression for the energy levels in the hydrogen atom. Years later, Bohr's assumptions were confirmed by the theory of quantum mechanics.

Colors and Charge

Does it make sense to assume that in nature only color singlet objects exist as real particles? Let us consider for a moment the comparable situation in electrodynamics. We suppose that color, which we interpret as some kind of charge, is similar to an electric charge. Everybody knows that objects with opposite charge attract each other and those with like charge repel each other. An electron and its antiparticle, the positron, attract each other and form a bound system, positronium, which we have already discussed (figure 9.4). The electron's charge is -1, that of the positron is $+1$, and so the sum of the charges (that is, the charge of the positronium) is zero. If we wish, we can call such a system a charge singlet state since it behaves like a singlet with respect to charge symmetry. Two electrons or two positrons, though, repel each other and so no bound state is formed. A two-electron system has a charge of -2 and a two-positron system has a charge of $+2$. Neither is a charge singlet. Thus the dynamics of electricity in nature re-

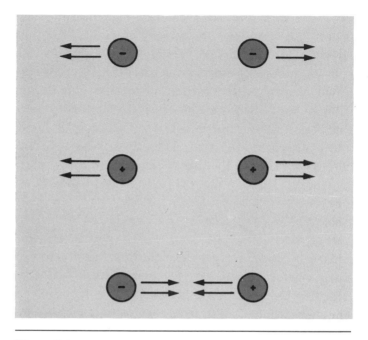

Figure 9.4
An electron and a positron attract each other and form a bound system, which may be interpreted as a charge singlet since the total electric charge of the system is zero. Two electrons or two positrons form a system of charge −2 or +2 in which the particles repel each other; thus, no bound system is formed.

semble the dynamics of color that we are looking for, at least in the sense that singlets under the corresponding symmetry (charge singlets versus color singlets) are the bound states of the underlying charged or colored particles. However, there is a big difference between electrodynamics and chromodynamics when it comes to the nature of their constituents (electrons versus quarks): electrons can be observed as independent particles, but quarks apparently do not exist as free particles.

Comparing electrodynamics with chromodynamics,

we notice a further important difference: three quarks can form a color singlet (namely, a baryon), but three electrons will form a state of electric charge −3, which is not a charge singlet. What is going on here? Well, in electrodynamics charge symmetry is a very simple matter. In a system with charge zero (charge singlet state), the number of positively charged particles must always equal the number of negatively charged particles.

Things are not so simple, however, in the color theory of quarks. Here we are dealing not with one charge, as in electrodynamics, but with several, and for this reason we cannot simply add and subtract charges. Adding up the color charges is more difficult; it is more like adding vectors in three-dimensional space than like counting numbers. We must know the mathematics of group theory to add color charges. Since we are not in a position to develop this mathematics here, we simply state the result: three quarks can indeed form a color singlet just as a quark and an antiquark can.

If quarks have this property of color (physicists often speak of "color quantum numbers" for quarks, which is nothing but a notation for their three color indices), it should be possible to find other manifestations of color. Indeed, physicists have found several indications that speak in favor of the existence of color. The first hint that the color quantum number is real came in 1971, when it was realized that the introduction of color can solve a long-standing puzzle. Using a special technique called current algebra, first introduced by Gell-Mann in the beginning of the 1960s, and provided we know the basic constituents of hadrons, we can calculate the lifetime of the neutral pion. As soon as the quark model was introduced, physicists calculated the lifetime of the neutral pion, which decays into two photons via the electromagnetic interaction, and came up with a result that was wrong by a factor of approximately 9. The lifetime of the neutral pion was calculated to be 7.5×10^{-16} s, but the observed lifetime turned out to be 0.83×10^{-16} s. Here it seems that we have one of the major disagreements between quark theory and the experiments in front of us.

Can the introduction of color quantum numbers change the situation? It can be shown that the lifetime of the neutral pion depends on the number of colors of the u and d quark flavors. The more colors present, the faster the pion decays. In the case of three colors, the pion would be expected to decay $3 \times 3 = 9$ times faster than if there were no colors, and on this basis we predict the lifetime of the neutral pion to be 0.83×10^{-16} s. This is precisely the lifetime observed, and so we conclude that, in accordance with our conclusions drawn from the analysis of the hadronic spectrum, there ought to exist precisely three different colors.

Another way of counting the number of colors is to investigate electron-positron annihilations that produce hadrons at high energies. When electrons and positrons annihilate via electromagnetic interaction, they produce either a pair of leptons or a number of hadrons. Since hadrons are supposedly composed of quarks, one expects that hadrons are produced via the production of a quark-antiquark pair turning into the final hadronic system (figure 9.5).

It is difficult to predict how often a particular hadronic configuration, a three pion system, for example,

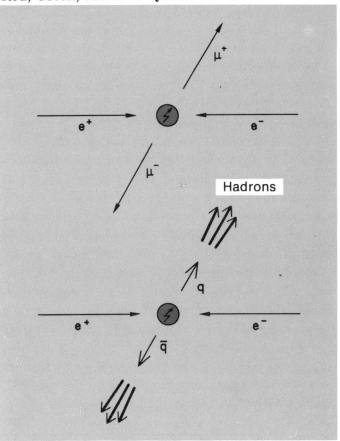

Figure 9.5
The annihilation of an electron and a positron into a pair of muons and a quark-antiquark pair. The production of a quark-antiquark pair finally leads to the production of a number of hadrons, which can be observed in the various particle detectors surrounding the annihilation region.

or a proton-antiproton pair, is produced in electron-positron annihilation at high energies. However, we can predict the overall production rate for hadrons; at least it is possible to do so as long as the total energy available in the annihilation process is large relative to that of the mass of a typical hadron. In such an instance, we expect the overall rate of the production of hadrons relative to the rate for producing a pair of muons to be given by the squares of the quark charges. This is so because the annihilation of an electron-positron pair and the subsequent production of a quark-antiquark pair is an electromagnetic interaction. The strength of any electromagnetic process, including the annihilation process under discussion here, is determined by the electric charges of the objects participating in the interaction. Therefore, the rate for producing hadrons via the initial production of quarks is fixed by the quark charges or, strictly speaking, by the squares of the quark charges.

Let us study the annihilation of electrons and positrons in the range between 2 and 3.5 GeV (the sum of the energies of the electron and the positron lies between these two values). In this region, we can produce only particles consisting of u, d, and s quarks; we cannot produce particles containing the c quarks. For this reason, we expect the following ratio:

$$R = \frac{\text{production rate for hadrons}}{\text{production rate for muon-antimuon pairs}}$$

$$= \frac{3(e_u^2 + e_d^2 + e_s^2)}{e_\mu^2} = 3[(\tfrac{2}{3})^2 + (-\tfrac{1}{3})^2 + (-\tfrac{1}{3})^2] = 2$$

The factor 3 stands for the number of colors. We have summed up the contributions of the three colored u quarks, the three colored d quarks, and the three colored s quarks, with the final result being that the probability of producing hadrons in electron-positron annihilations in the range between 2 and 3.5 GeV is twice as high as the probability of producing muons.

Red, Green, and Blue Quarks

Color Is Verified

Experiments have verified that this predicted ratio is indeed close to 2, about 2.1 or 2.2. We emphasize that without the color quantum number this ratio would be three times smaller (that is, 2/3), which does not agree at all with the experiments.

We therefore conclude that there exist three different phenomena in nature to support the existence of the color quantum number:

1. The structure of the hadronic spectrum
2. The lifetime of the neutral pion
3. The rate of hadron production in electron-positron annihilation

So far, however, we have not mentioned the ultimate success of color theory, and this we shall discuss in the next chapter. As we shall see, the introduction of the color quantum number allows the formulation of a new theory of hadrons and of the strong interactions: quantum chromodynamics. Today we believe that this is the correct theory of hadrons, which physicists have been looking for since the 1930s.

.

X

A Theory of
Hadrons Called QCD

As mentioned in the previous chapter, there is a crucial difference between electrodynamics and the quark model in that three electrons can never form a bound state but three quarks apparently can do so with ease. What force explains this odd phenomenon? It has to be some force that attracts three quarks in such a way that a baryon is formed.

Let us first of all state one important conclusion about the nature of interquark forces. Consider the following two configurations of three quarks: a typical baryon configuration composed of a red, a green, and a blue quark and a configuration of two red quarks and one blue one. The first configuration, which involves all three colors, can be in a color singlet state; the second configuration lacks the green quark and therefore can never be in a color singlet state.

A Theory of Hadrons Called QCD

Now let us suppose that the interquark force acts with the same strength on all three colors, in other words, that it does not discriminate between the various colors and is, as it were, color-blind. If that is the case, however, the force will make no distinction between our two configurations and this would imply that they require the same amount of energy. In other words, color singlets and nonsinglets would be indistinguishable from each other. But this is precisely what we do not want to happen. Therefore, things must be arranged in such a way that color singlets and nonsinglets differ in the amount of energy they have. The nonsinglets should have the higher energy or, preferably, should not even exist. They only way to arrange for this is to introduce a color-dependent force that acts with different strength on each of the three colors.

Let us recall once more what happens in electrodynamics. An electrodynamic field creates the force of attraction between two charged objects having opposite charge. In the language of quantum theory, the force is created through the exchange of virtual photons between the charged objects. These photons couple to the charged objects with a strength described by the electric charge.

Let us suppose for the moment that the forces between quarks are similar to the forces between electrons and positrons in the sense that we can describe the quark forces by the exchange of virtual quanta, similar to the virtual photons. These virtual quanta we can call gluons. Let us further suppose that the coupling of these gluons to quarks is proportional to the color charges of the quarks. In this way, we are constructing a kind of electrodynamics of color space in which the color charges assume the role of the electric charge and virtual photons are replaced by gluons.

The coupling of gluons to quarks is more complicated than the coupling of photons to electrons, however. Whenever a photon interacts with an electron, the latter remains an electron, but a gluon interacting with a quark can change the color of the quark. For example, a gluon can transform a red quark into a green quark (figure 10.1).

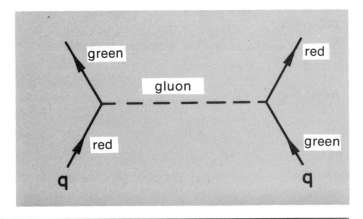

Figure 10.1
The interaction between gluons and quarks. The colors of quarks can change when they interact with gluons.

The gluons can be labeled according to their color-carrying properties. Three different colors allow nine different ways of coupling gluons to quarks:

red	→	green	blue	→	green
red	→	blue	red	→	red
green	→	red	green	→	green
green	→	blue	blue	→	blue
blue	→	red			

Note that the last three couplings differ from the others in that the colors do not change. In particular, there is a situation that is completely symmetric with regard to color; namely the superposition red $\longrightarrow$ red + green $\longrightarrow$ green + blue $\longrightarrow$ blue. We disregard this type of coupling since it can produce no change of colors.* Thus, only two independent superpositions of the last three couplings count. Together with the first six we are dealing with a total of eight different couplings. We now proceed to assume that there is a gluon for each of them, that is, we suppose eight different gluons.

The Color Analog of QED

We are now in a position to examine in greater detail the forces between quarks and the energies of the various quark states. Even though the outcome is exceedingly simple, however, we shall dispense with the actual mathematical calculations because they involve a familiarity with group theory that I do not wish to presuppose. At any rate, we find that color singlets are the bound states of our theory just as neutral charge singlets are the bound states of electrodynamics. Quarks, therefore, tend to cluster in color singlet configurations if the forces between them are due to the exchange of gluons. Keeping this result in mind, we now ask whether we can construct a real theory of quarks and gluons that will incorporate these features? For electric forces, we have Maxwell's

---

* Mathematicians would say that this type of coupling "remains a singlet under the color group SU(3)."

theory and its quantitized version, quantum electro-dynamics. We are searching for the color analog of quantum electrodynamics, which we shall call quantum chromodynamics.

To develop these ideas just a bit further, we suppose the theory of quarks and gluons to be merely a modified form of electrodynamics with color replacing the electric charge. One way of describing this theory is to compile a dictionary in which the various notions of quantum electrodynamics (QED) are translated into notions of quantum chromodynamics (QCD):

QED	QCD
electron	u quark
myon	d quark
. . .	. . .
electric charge	color charge
photon	gluon
atom	meson, baryon

Looking at this dictionary may lead us to think that QED and QCD are entirely analogous. This is not the case, however. Electrodynamics has only one charge, the usual electric charge, whereas chromodynamics has eight different color charges. Using the language of mathematical physics, we say that the color charges form a group, the color group SU(3). This group is called the gauge group of chromodynamics, and for this reason the theory of chromodynamics is called a non-Abelian gauge theory (because the color group SU(3) is an example of what is called a non-Abelian group). Theories of this kind were first studied by the German mathematician Hermann Weyl just after World War I and were later further

A Theory of Hadrons Called QCD

developed by Weyl and by the Swedish physicist Oscar Klein. Non-Abelean gauge theories were then explored in more detail by the American theorists C. N. Young and R. Mills in 1953.

One of the most important properties of quantum chromodynamics is that gluons carry color. They have the same color properties as color charges, that is, they are octets with respect to color symmetry. The eight gluons can be described by their color-carrying properties. Above we described the various ways in which gluons can change the color of a quark: red $\rightarrow$ green, and so on. In figure 10.2, we describe

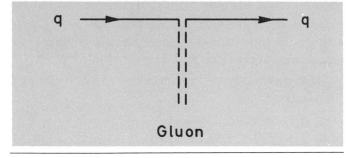

Figure 10.2
A quark-gluon vertex. The gluon is denoted by the dashed double line.

such a quark-gluon coupling (called a quark-gluon vertex). Gluons can be thought of as lines carrying color indices. These color lines show that the gluons can transform a quark of one color into one of a different color.

As we shall see later, it is of crucial importance to the physics of the strong interaction that the gluons in QCD have color. First of all, though, let us note that this creates a situation very different from the one

143

that obtains in electrodynamics, where the charged particles are always fermions: that is, spin-1/2 particles, electrons, positrons, protons, and so on. The photon that generates the electric forces between the charged particles is itself neutral. That the photon does not carry an electric charge is a very important fact for the physics of the electrodynamic interaction. A neutral photon implies, for example, that in the absence of matter composed of charged particles, electromagnetic waves (laser beams, for instance) can propagate freely through space. Nothing disturbs them. This would not be so if photons had an electric charge. For if photons were charged, they would interact with each other, and a phenomenon as simple as a ray of light would suddenly be something quite messy.

Since gluons have color, there is interaction not

Figure 10.3
The interaction of three gluons in QCD (a three-gluon vertex). The various color lines flow continuously from one gluon to the next.

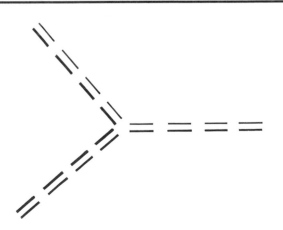

only between gluons and quarks but also among gluons. In particular, there is a vertex among three different gluons. Figure 10.3 shows three gluons meeting at a point. Notice that each color line is smooth and unbroken from gluon to gluon. No color line ends at any point, and none starts fresh. For example, we cannot have a three-gluon vertex where a red-green and a red-blue gluon meet to produce a red-green gluon because this would mean an end to the blue color entering from the left in figure 10.3 and the start of another red color. This the rules of the game (the mathematical rules of QCD as a special example of non-Abelian gauge theory) do not allow.

The Vacuum Is Not Empty

Let us investigate the quantum properties of chromodynamics a bit more closely. Again we first look at the corresponding situation in electrodynamics. We mentioned above that electromagnetic waves propogate freely through space if they are not disturbed by charged particles. However, this holds true only for classical physics. A vacuum, which looks so simple to macroscopic observers like us, is actually a very complicated system in quantum theory. The uncertainty relations of quantum mechanics have it that the precise measurement of the momentum or the energy of a particle requires either considerable space or considerable time. If we wish to explore the structure of objects at very small distances—less than 10^{-12} cm, for instance—it becomes impossible to determine the momenta of a particle with any degree of accuracy. On the contrary, the uncertainty in the momenta of particles becomes considerable.

Something very odd happens in electrodynamics once the uncertainty in energy is more than twice the mass of the electron (as occurs at a distance of about 10^{-11} cm): there a pair of particles consisting of one electron and one positron can be produced. This particle pair is produced out of the vacuum, and, unless something drastic happens, it will disappear just as quickly. This something drastic might be a supply of energy from outside the vacuum so that the electron-positron pair becomes a pair of genuine particles, one that does not violate the law of conservation of energy. Without such an outside supply of energy, the electron-positron pair would exist not as a pair of real particles but merely as what we call virtual particles. The vacuum is filled with many, (essentially infinitely many) virtual electrons-positron pairs. They influence the physics at exceedingly short distances, but play no role in macroscopic physics.

We mentioned virtual particles earlier in our discussion of physics. Recall that the force between electrically charged particles is generated by the exchange of virtual photons between them. Now we learn that there are not only virtual photons but also virtual electrons and positrons. In general, we can define a virtual particle as a would-be particle. It has no definite mass and exists for only a short time and in a tiny region of space. Uncertainty relations are responsible for the appearance of virtual particles in physics.

What was just said about electron-positron pairs also holds true for muon pairs and even for quark-antiquark pairs. For example, the production of a muon-antimuon pair in electron-positron annihiliation can be understood in the following manner. The vacuum is filled with virtual muons and antimuons. Normally we do not see them. However, in the annihilation of an electron and a positron—in a colliding-beam experiment, for example—we supply energy to a very small region in space. A virtual muon-antimuon pair in that region jumps on this opportunity to become a pair of real muons, which leaves the region of interaction as free particles and can be observed in particle detectors.

Let us suppose we measure the charge of an electron by studying the scattering of two electrons in a laboratory. The two electrons repel each other, and the repelling force between them, according to Coulomb's law, is proportional to the square of the electric charge of the electron. Normally two electrons do not come very close to each other when they are scattered in this manner. The distance between two electrons is always much greater than the typical lengths of quantum electrodynamics (about 10^{-11} cm), and in such cases Coulomb's law quite accurately describes the force between the two electrons.

The Breakdown of Coulomb's Law

What happens, however, if we try to measure the force between two electrons at distances smaller than 10^{-11} cm (which is done experimentally by increasing the energies of the scattering electrons)? It turns out that Coulomb's law does not hold once two electrons come closer than 10^{-11} cm to each other. At such close distances, the forces between the electrons are a bit larger than expected on the basis of Coulomb's law. How is this effect to be interpreted?

What we observe here is the effect of the infinitely extended sea of virtual electron-positron pairs that fill the space between the two electrons. Suppose we take a single electron and set it free in space, that is, in the vacuum. Well, naively it might be expected that nothing much happens in such an event, that the electron just sits there. But no, matters are not that simple, not if we take into account that the vacuum is not really a vacuum but is filled with those invisible virtual electron-positron pairs. A change then occurs in the vacuum surrounding the electron. Since the electron is negatively charged, it repels all virtual electrons in its neighborhood and attracts the virtual positrons. Now the electron is surrounded by a cloud of virtual positrons and we say that the vacuum is polarized by the electron (figure 10.4). What is the net effect of this? If a cloud of virtual positrons surrounds the electron, the electric charge of the electron is shielded to a certain extent. If we look at the electron from a great enough distance for the vacuum polarization effects to be negligible, we see the electron and its cloud of virtual particles as a whole and cannot distinguish which part of the electric charge derives from the electron itself and which part from the cloud. This is what physicists call a physical electron: an electron and its vacuum polarization cloud. An electron without a vacuum polarization cloud is called a naked electron.

The reader may now be confused and ask, What is

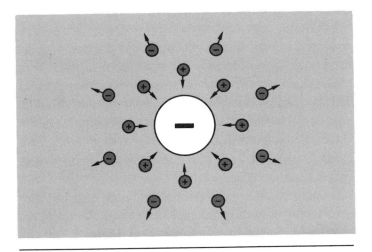

Figure 10.4
The effect of vacuum polarization in electrodynamics. An electron surrounds itself with a cloud of virtual positrons. The electric charge of the electron is thus partially shielded.

the actual electron we use for experiments? What transports electricity in wires? The answer is, It depends. In processes where the energy of the electron is small relative to its mass, we cannot explicitly see the effects of vacuum polarization. This is what happens in most everyday instances. In these cases we are dealing with the "dressed" physical electron. However, if the energies involved are sufficiently large (in the scattering process, for example), we can see the vacuum polarization effects explicitly. We begin to undress the electron, as it were. Since the electric charge of the naked electron is greater than that of the physical electron, the force (relative to Coulomb's law) with which the two approaching electrons repel each other increases. This explains the deviation from Coulomb's law mentioned above.

QED Is 'More Than Perfect'

Thus far we have made only qualitative speculations, and the reader may ask whether it is possible to make exact quantitative predictions for processes in which vacuum polarization effects are important. The answer is yes, and the theory that allows us to make such calculations is quantum electrodynamics. This theory works not only to our satisfaction but actually better than we expected it to. Very precise tests have been made of quantum electrodynamics, and the theory cannot be wrong by more than one point in one million. Therefore we can say that there is nothing mysterious about the interaction of electrons and photons. The electromagnetic interaction is completely comprehensible not only within the terms of classical physics, where everything is described by Maxwell's equation, but also when we take quantum phenomena into account.

It is good to know that something really works. It confirms that research in physics is on the right track toward an understanding of the vacuum polarization effect.

Normally the vacuum is filled not only with virtual electrons, positrons, and photons (which account for vacuum polarization in electrodynamics) but also with virtual quarks, antiquarks, and gluons. So what happens when we place a quark in a vacuum, which we shall pretend to do despite the fact that quarks are not supposed to lead an independent existence as real particles. For the sake of our hypothesis, as a kind of thought experiment, we lend our quarks a hypothetical existence, as electrons, for example, and study the consequences of our hypothesis.

We expect that a quark in a vacuum will behave pretty much as an electron would under the same circumstances: we expect it to polarize the vacuum. Acting with their color charge, the quarks should attract other nearby antiquarks while simultaneously repelling other quarks that are about. All we need to do is replace (1) the electric charge in QED with the color charge in QCD and (2) the virtual electron-positron pairs in the vacuum with virtual quark-antiquark pairs. The net effect in QCD should be the same as in QED. The color charge is partially screened, and the naked quark we had at the beginning is in a cloak of virtual antiquarks. This we would call a physical quark if there really were such an object. QCD and QED are really quite analogous in this respect. We do not even need to draw a new picture. If the reader wishes to see how a quark behaves in a sea of virtual quark-antiquark pairs in the vacuum, she or he is invited to look at figure 10.4 and to think of the circles as quarks.

Is this really all that happens in QCD, though? Absolutely not. We have forgotten the most important effect. The vacuum is filled not only with virtual quark-antiquark pairs but also with gluons. Recall that something similar obtains in electrodynamics— there the vacuum is filled not only with virtual electron-positron pairs but also with virtual photons. Why did we not mention the photons in our discussion of the vacuum polarization effects in electrodynamics? Because there they play essentially no role since they are electrically neutral. By placing an electron in the vacuum, we influence only the sea of virtual electron-positron pairs, not the sea of virtual photons. However, things cannot work quite the same way in chromo-

dynamics because gluons carry a color charge just as
quarks do. Thus, if we place a colored object like a
quark in the vacuum, we are going to exert an influence on the gluon sea. A quark will surround itself
not only with a cloud of virtual antiquarks but also
with a cloud of virtual gluons. The cloak of a quark is
more complicated than that of an electron.

The Explosion of the Vacuum

One might suppose that this is all that happens,
that in other respects the vacuum polarization in
QCD is quite analogous to that in QED, that is, the
dressed quark has a net color charge that is less than
the color charge of a naked quark. This, of course, is
just a guess, and we must calculate precisely to see
what really transpires in QCD. The calculations are
far too complicated for this book, however. The result, surprisingly, is the opposite of what we expect. It
turns out that the cloud of virtual gluons surrounding
the quark does not diminish the quark's color charge;
it actually increases it.

To explain this surprising phenomenon, we invent a
model that resembles electrodynamics stood on its
head. This topsy-turvy electrodynamics has electrons
and positrons and photons as does the normal world.
The only difference is that opposite charges repel
each other and like charges attract each other! What
happens in such a world when we place an electron in
a vacuum? It surrounds itself with a cloud of virtual
electrons (and not with positrons, as an electron in
the real world does), and this increases the electric

A Theory of Hadrons Called QCD

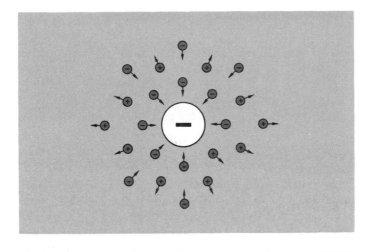

Figure 10.5
The vacuum polarization in modified electrodynamics. Equal charges attract each other; unequal charges repel each other. An electron surrounds itself with a cloud of virtual electrons, and this increases the effective electric charge of the electron and influences the vicinity of the electron even more strongly. An unstable situation is the result.

charge due to vacuum polarization (figure 10.5). Now the electron plus its electron cloud have an even stronger effect on the surrounding cloud of virtual electron-positron pairs. More electrons are attracted, more positrons repelled. We realize at once that this is a process with no end. The more virtual electrons are "consumed" by the cloud surrounding the electron, the stronger the force on the remaining electrons. The vacuum polarization cloud becomes larger and larger; it grows like a cancer, devouring the virtual electrons in the vacuum. We arrive at a somewhat uncomfortable state of affairs, a kind of explosion, and in this highly unstable situation begin to appreciate the balanced world of real electrodynamics. When an electron is placed in a vacuum in the real world, after all, there is only a slight reorganization of the sea of virtual particles, which quickly stabilizes itself.

How are we to interpret the hypothetical events in this altered electrodynamic world? The instability is caused by charged particles and thus would be eliminated if there were no charged particles, only neutral particles—in other words, if all charged particles were confined.

There is no further use in studying this topsy-turvy electrodynamics. The only reason it was brought up was because in QCD the gluon cloud surrounding a quark behaves like an electron cloud in an electrodynamic world that has been stood on its head. Gluons surround a quark and increase its color charge with the consequence of increasing the disturbance of the surrounding vacuum. More and more gluons are attracted. The gluon cloud "consumes" gluons and increases without limit. A quark in chromodynamics behaves something like a voracious beast whose appetite increases the more it eats. Here, too, we have a highly unstable situation.

Such a beast will soon eat itself to death.

Quarks and Gluons Are Confined

The simplest way to avoid the predicament of instability in QCD is to assume that there are no isolated quarks, that they exist solely in color singlet configurations like mesons or baryons. Of course, this is precisely what we want.

Is it really true, however, that in QCD colored objects (quarks, gluons) cannot exist as free particles? Our considerations suggest this very strongly but do not constitute definite proof. Right now there is no proof for the absolute confinement of color. However, theory and experiment so far indicate that color is confined and that the quarks and gluons can appear only in "white" states. For our subsequent considerations, we shall simply assume that this is so.

XI

Chromoelectric Confinement of Color

The best way to discover the cause of the instability of the vacuum is to calculate the behavior of the gluon cloud in QCD, using the rules of quantum mechanics. To do so, we encounter a remarkable phenomenon. Before discussing this phenomenon in detail, however, let us analyze what occurs in the analogous situation in electrodynamics. The photon quanta responsible for the generation of the forces in QED come in two varieties. One, the virtual photons, which produce the electric attraction between electrons and protons, we call the coulombic quanta. These exist only in the presence of charged particles and never in isolation as real photons. The second variety are called transverse or magnetic quanta. These photons may exist as real particles, as constituents of a laser beam, for example, and produce magnetic forces between charged particles.

Color, Magnetism, and Electricity

Matters are quite similar in QCD, and if we want to we can classify all gluon quanta as either color coulombic or color magnetic quanta. Looking at vacuum polarization in detail, we discover that the instability of the gluon cloud in QCD is caused by a rather ingenious interplay of these two kinds of quanta. The point is that the color coulombic quanta and the color magnetic quanta interact with each other, which is impossible in QED because photons lack charge. For example, a color coulombic quantum can split into a color coulombic quantum and a color magnetic quantum, and these two quanta can then recombine into a coulombic quantum (figure 11.1). It is this interac-

Figure 11.1
An important phenomenon in QCD: a colored coulombic quantum (dashed line) splits into coulombic and magnetic quanta. Subsequently, the two quanta recombine into a coulombic quantum.

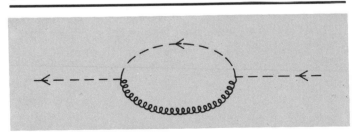

tion between two kinds of gluon quanta that leads to the instability of the vacuum in QCD.

We can illustrate the importance of this interaction in a slightly different way. QCD has the unique feature that the charge of a colored object—a quark, for instance—increases if we add gluonic vacuum polarization effects. The vacuum instability, however, occurs only at relatively large distances (distances greater than the diameter of the proton, which is about 10^{-13} cm). At such large distances the color charge becomes so great that it obviates our previous methods of calculation (that is, the methods we used to calculate vacuum polarization effects in QED). If we regard quarks and gluons within a confine narrower than 10^{-13} cm, however, the problem lessens and we can use the analogy of QED in analyzing the effects of vacuum polarization in QCD.

Realizing this, we can use the following trick. Let us consider a very heavy quark and its antiquark, which form a quark-antiquark bound state, a heavy meson. Heavy in this case means relative to the mass of the proton (about 940 MeV). For example, let us take a quark-antiquark system with a mass of 50 000 MeV. Of course, we do not know whether such quarks exist in the real world (though we might in a few years), but never mind. Even if they do not exist, we can conduct our thought experiment. The constituents of such a heavy meson system are very near each other; the diameter of this system is on the order of 10^{-15} cm, fifty times smaller than a proton.

Chromoelectric Confinement of Color

The Color Analog of Coulomb's Law

If quarks are this close to each other, we can ignore the effects of the gluonic vacuum polarization cloud. In this instance, the force on the quarks can be described by Coulomb's law—heavy mesons are very similiar to electron-positron bound states in QED or to the hydrogen atom. Specifically, the energy levels of such systems will resemble those of the hydrogen atom, which the reader may recall from high school physics. Of course, we do not know if this is really the case in Nature. It would certainly be fascinating if it were, though, for then we could test the color coulombic forces simply by looking at the spectrum of hadrons.

However, this picture of two heavy quarks bound by the color coulombic force is not entirely accurate. The gluons surrounding the quarks do play a certain role, and one way of describing the coulombic force between two quarks is to draw chromoelectric field lines between them (figure 11.2). The thinner these field lines, the weaker the corresponding color force. If we pull the quarks apart, the field lines become thinner and the force between the two quarks decreases as the square of the distance separating them.

Let us now consider the effects due to the quantum fluctuations between gluons. This net effect can be described very simply by recalling that in QCD there is an interaction between color coulombic quanta and color magnetic quanta. This interaction causes a mutual attraction of chromoelectric field lines, an attraction produced by magnetic quanta "running" between the field lines. This is indicated in figure 11.2 by the wavy lines running from one field line to the next.

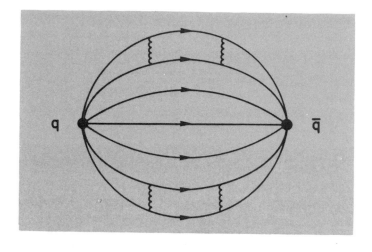

Figure 11.2
The chromoelectric field lines between a heavy quark and its anti-quark. Note that the chromoelectric field lines influence each other by exchanging color magnetic quanta. This leads to a compression of the chromoelectric field lines, a unique phenomenon of QCD that does not exist in electrodynamics.

This attraction of chromoelectric field lines, which can be precisely calculated if quarks are close to each other, makes the field lines thicker than they would be if the lines running between quarks were described exclusively by Coulomb's law. This, of course, is just what we expected on the basis of our previous considerations. The color charges of quarks increase as the distance between them increases, and we encounter a force that is slightly greater than what we would expect on the basis of Coulomb's law. We emphasize once more that the crucial new aspect of QCD is the chromomagnetic attraction of the chromoelectric field lines. This attraction results from the fact that gluons carry color, and this also explains why such an effect cannot be observed in electrodynamics.

160

Chromoelectric Confinement of Color

Pulling Quarks Apart

Let us now turn back to our heavy meson system. We already know that, as we try to pull two quarks apart, the force between them decreases approximately as the square of the distance between them. We also know that Coulomb's law does not describe the force between the two quarks with complete precision, that the force is a bit stronger because we are dealing with gluons. This increase is a result of the vacuum polarization effect of the virtual gluons. Now let us see what happens if we pull the two quarks apart. For a time, the interquark forces keep decreasing, approximating Coulomb's law. However, the vacuum polarization effect simultaneously increases the color charges between the quarks, and so the chromoelectric field lines between the quark become more and more compressed. What ultimately happens once we pull the two quarks very far apart, for example, to the macroscopic distance of 1 m?

Unfortunately I have to confess that we do not precisely know what happens in such an instance. We can make precise calculations in QCD only when quarks are very near each other, much closer than 10^{-13} cm. So in the absence of precise information all we can do is speculate about what happens at macroscopic distances. First of all, we must emphasize that it will not be easy to separate the quarks by 1 m. Yet if we could, we would be able to study them in isolation, measure their electric charge, and so on. Separating a quark-antiquark system by a distance of 1 m is nearly the same thing as producing free quarks, which we know is either exceedingly difficult or impossible.

Most physicists today believe that it is at least theoretically possible to separate quarks in a meson system up to macroscopic distances, but only at the price
of introducing large amounts of energy. It is believed
that chromoelectric field lines running from one
quark to the other become increasingly compressed as
we try to pull the quarks apart and that we finally
arrive at a configuration where the field lines run parallel to each other, like the field lines between two
condenser plates in electrodynamics (figure 11.3). It

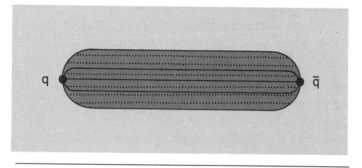

Figure 11.3
If quarks are separated by more than 10^{-13} cm, the chromoelectric field lines become parallel. The force between quarks at such
distances is independent of the distance between them, just like
the force between condenser plates in electrodynamics.

is expected that the field lines become parallel once
quarks are more than 10^{-13} cm apart. Now the force
between condenser plates in electrodynamics is independent of the distance between the plates, and the
same is expected to hold true for the force between
quarks. At large distances, the interquark force is believed to be constant. Even if we succeed in separating the quarks in a meson system up to the distance
of 1 m, the force between them will then be just as
strong as it was when the separation distance was

Chromoelectric Confinement of Color

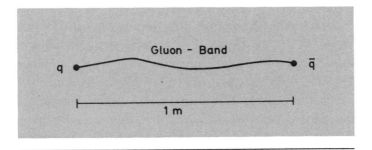

Figure 11.4

A quark and its antiquark separated by a macroscopic distance. The strong force between the two particles is generated by the thin strings of gluonic field lines (thickness about 10^{-13} cm).

anything greater than about 10^{-13} cm. It is the chromoelectric gluon field lines (figure 11.4) that produce the strong force and connect the quarks.

How strong is the force that acts between the quarks? I have already mentioned that until now we have been unable to calculate this force from first principles of chromodynamics.* However, we can do something else. We can take a look at the relatively familiar spectrum of the hadrons, and from the spectrum of quark-antiquark systems like the mesons we can say something about the force between quarks. The situation once again is analogous to the one that obtains in electrodynamics. If we study the energy

---

*Until very recently, it was not possible to calculate the force between quarks at large distances from the equations of QCD. However, in 1981 and 1982 theoretical physicists made considerable progress in the understanding of quantum chromodynamics by using a numerical method called lattice approach, developed by Kenneth G. Wilson of Cornell University and others. The idea is to replace the space-time continuum by a lattice of space-time points. In this way, it is possible to use large computers to solve the equations of QCD. The results, although still preliminary, support the picture of the chromoelectric confinement of color described here.

levels of the hydrogen atom, we can estimate the force between the proton and the electron inside the atom. What we find is, of course, the well-known force between charged objects given by Coulomb's law, which decreases as the square of the distance between the objects. If we apply the same principle to the meson spectrum, we find that the force between quarks at relatively large distances (larger than 10^{-14} cm) is constant and indeed independent of distance, which is precisely what we used above.

The force between quarks is very strong indeed. It is so strong, as a matter of fact, that we need an enormous amount of energy—as much as is required to lift 1 ton of matter by 1 m—to separate two quarks by a distance of 1 m. This energy is so great that we shall never be able to separate quarks by macroscopic distances, and of course it is far stronger than the electric forces responsible for the structure of the atom. It is also stronger than the force responsible for the structure of atomic nuclei. In studying the structure of hadrons, we have encountered the strongest known forces in nature, the chromoelectric forces between quarks.

Some readers may wonder why the energy mentioned here is so enormous, since after all machines lift objects weighing one ton every day. The point is that we must concentrate this energy on a single quark-antiquark pair. The most powerful accelerators physicists are using today are able to separate quarks from each other by distances of about 10^{-11} cm. In order to pull quarks away from each other by one meter, one would need accelerators 10^{13} times more powerful. Even an accelerator which is built around the equator would by far not be sufficient to produce the required energy to separate quarks from each other by macroscopic distances.

Once again we mention the fact that color charges between colored objects decrease at small distances but increase at large distances. Until now we have studied only the strengthening of the force at large distances and have suggested that this leads to a permanent binding of quarks inside hadrons. What about the related weakening of the force at small distances, though? Recall that in an earlier chapter we discussed the outcome of the SLAC electron-scattering experiments showing that quarks seem to be essentially structureless. We concluded that the strong force between quarks must become weak at small distances. The strong interaction dies off. How gratifying that QCD has this property, for it is the only theory that can explain what happens in the SLAC experiments. This provides another strong reason for supposing that QCD is the correct theory of hadrons.

Asymptotic Freedom and Infrared Slavery

That the force between quarks becomes weak at very short distances is called asymptotic freedom. At such distances, quarks behave like independent particles and no strong force exists between them. However, we also found that quarks bind together because of that other fundamental characteristic of QCD, the increase of color charge at large distances.

In the language of wave mechanics, one often uses the term "infrared" for "long distance"; and for this reason the phenomenon of the strengthening of the color charge at large distances is called infrared slavery. In a way, the quarks are slaves of their own color charge. If we wish, we can think of them as bound like prisoners on a chain gang. If the prisoners stay within a few meters of each other, nothing much happens. They can move more or less freely. For example, they can walk around and no one may even notice anything unusual if the chains are kept out of sight. Thus we can say that our prisoners are free within limits. The problem begins only when one prisoner tries to get more than a few meters away from the others. If that happens, all the prisoners are again reminded that they are in chains.

Replacing the prisoners by quarks and the chains by chromoelectric strings, we arrive at an analogy that we believe accurately describes the dynamics of quarks and gluons inside a hadron. However, there is one essential difference between our chain gang example and quarks. Any locksmith can break the chain between two prisoners, but no locksmith is expert enough to break the gluon chains between quarks. Quarks remain slaves forever.

What Are the Nuclear Forces?

The original impetus of particle physics was the desire to understand how atomic nuclei function. What holds a proton and neutron together to form the object we call a deuteron? What holds six protons and six neutrons together to form a carbon nucleus? In searching for the answer, we have arrived at the physics of quarks and gluons, and we have an approximate idea of what holds quarks together to form hadrons. But what holds a proton and neutron together to form a nucleus? It seems that in all the excitement over quarks we have forgotten the original question that led to the discovery of the quark structure of hadrons in the first place. With our newly acquired knowledge about quarks, we shall pose the question again: what is the nature of the nuclear forces?

Before trying to answer this question, let us return once again to electrodynamics. The electrons and protons inside an atom attract each other to form an electrically neutral atom. Since atoms have no electric charge, they do not influence each other electrically. In other words, there is no attraction or repulsion between atoms. This is not strictly true, however, unless atoms are comparatively far away from each other, far, that is, relative to the diameter of an atom, which is of the order of 10^{-8} cm. If atoms are fairly close, there will be a slight electric interaction between them. This force is generated because atoms have a finite size and the charges inside an atom are not concentrated at one point but distributed throughout. Specifically, if two atoms come very near each other, the electrons in the atomic clouds will produce a strong repulsion between the two atoms. The existence of such forces between atoms was first pointed out in the last century by the Dutch physicist Johannes van der Waals, whom these forces are named after. What do these van der Waals forces have to do with nuclear forces? Well, it is thought that nuclear forces are nothing but the van der Waals forces of QCD. Hadrons are color singlet configurations, and for this reason the superstrong color forces do not affect them, just as the electric force does not act directly on atoms. However, the color forces between quarks can act indirectly to produce the color analog to van der Waals forces between hadrons. Like atomic van der Waals forces, the QCD van der Waals forces should also disappear quickly once the distance between hadrons is increased to more than 10^{-13} cm, and this is precisely what we observe. Nuclear forces act only at small distances and disappear once the distance between two nucleons is larger than 10^{-13} cm.

The emerging picture of nuclear forces is one that nobody had anticipated. Until recently, most physicists were convinced that the nuclear forces were fundamental and that nothing could be stronger. Today the situation has changed completely. The nuclear forces are simply the color van der Waals forces between hadrons. They are not fundamental. Rather, like atomic van der Waals forces, they are complex phenomena that depend on the structure of the nucleons. Moreover, the nuclear forces are merely the detritus of the huge color forces between quarks. The situation we have arrived at is indeed rather remarkable. A few decades ago, physicists set out to find the nature of the strong nuclear force, and instead they encountered a whole new world: the world of quarks and gluons and chromodynamic forces inside the nucleons.

XII

Chromomagnetic Forces

So far we have mainly discussed the chromoelectric forces between quarks, that is, the forces responsible for their binding. However, in QCD we also have chromomagnetic forces, which are the color analog of the magnetic forces in QED. As we have mentioned, the chromomagnetic effects are responsible for the confining characteristic in QCD. Is it possible to observe them directly, however?

To answer this question, we return once more to electrodynamics. Of course, we all know the importance of magnetic forces in our daily lives. Without them we would have no electric power plants, no electric engines, and on and on. This magnetic force also plays an important role inside the atom.

Magnetic Forces Between Electrons

The main force acting between electrons inside the atomic cloud and nucleons in the atomic nucleus is electric attraction. However, there is also a small magnetic force because the electron and the atomic nucleus both have spin. The spin of a particle is a kind of angular momentum, and since the angular momentum of a charged particle implies that electric charge moves around in circles, a charged particle with nonzero spin will generate a magnetic force. The magnetic field of a charged particle is similar to the magnetic field of a bar magnet (figure 12.1).

Two electrons whose spins point in the same direction attract each other magnetically; two electrons with opposite spins repel each other. However, compared to the electric interaction between electrons,

Figure 12.1
The magnetic field of an electron is similar to that of a bar magnet. The spin of the electron is denoted by the white arrow.

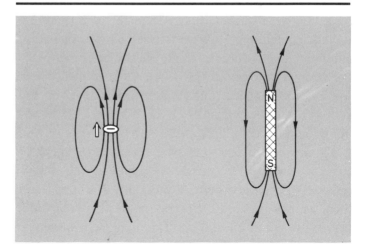

the magnetic force between them is very small. For example, the magnetic force between the electron and proton in a hydrogen atom is minute relative to the electric attraction of these two particles. The effect of this difference in the strength of the two forces is that hydrogen atoms with different spins have slightly different energy levels since the energy in a state where the spins of the electron are parallel differs from the energy in one where they are not. This slight difference in energy levels is called hyperfine splitting (a name that indicates we are dealing with a fractional effect).

An Exercise in Magnetics

Let us study another case, namely, the magnetic force between an electron and a positron. Since a positron has positive charge, the situation here is the exact opposite of the one described above. If the spins of the electron and positron point in the same direction, we observe magnetic repulsion between them; if the spins point in opposite directions, the forces attract each other. This is important for the structure of positronium, the bound state of and electron and positron.

We know that there are two kinds of positronium, orthopositronium and parapositronium. Orthopositronium consists of a positron and an electron whose spins point in the same direction (that is, the total spin is +1), and parapositronium consists of an electron and a positron whose spins are antiparallel. Therefore, we can produce parapositronium out of orthopositronium by simply reversing the spin of one of the constituent particles. Since the magnetic forces in the para and ortho cases are different in sign, we expect the two states to have slightly different levels of energy. In orthopositronium the attraction of the electric forces is counteracted by the magnetic repulsion. In parapositronium electric and magnetic forces both attract, and it is the magnetic field that helps form the parastate. In orthopositronium we therefore have to work against the magnetic force, whereas in parapositronium we do not. For this reason, we expect orthopositronium to have slightly more energy than parapositronium, which is precisely what has been observed experimentally.

Heavy Quarks Ignore Magnetism

What parallel do we expect to be the case in QCD? Let us first regard the mesons, the QCD analog of positronium. The chromomagnetic forces between two quarks are completely analogous to the magnetic forces in positronium. We therefore expect mesons whose two quark spins are aligned (an "orthopositronium meson," as it were) to be heavier than a meson whose two quark spins are opposite each other (as they are in parapositronium). And this is precisely what we observe. For example, the ρ meson is heavier than the π meson, the K* meson is heavier than the K meson, and so on. We find that the rule applies for all the mesons.

We bring up one further case, the system of c quarks and antiquarks. The J/Ψ meson is heavier than the η_c^- meson since in the J/Ψ the spins are aligned, and in the η_c they are opposed. The mass difference between the ρ meson and the π mesons is fairly large, about 600 MeV, but that between the J/Ψ meson and the η_c^- meson is much smaller, only about 100 MeV. Why is that?

It turns out that this phenomenon can easily be understood in QCD. The magnetic and the chromomagnetic forces between quarks are caused by what are called the magnetic moments of the quarks. (The magnetic moment of a particle defines how strongly the particle interacts with a given magnetic field.) The magnetic moments of electrons, muons, and, to some extent, quarks can be calculated and are directly proportional to Planck's constant h but inversely proportional to the mass of the particle. In other words, the larger the effective mass of a quark, the smaller its magnetic moment. Heavy quarks thus have less magnetic and chromomagnetic interacting capacity than lighter ones. For a c quark, the chromomagnetic force is much smaller than for a u or d quark. For this reason the difference in mass between the J/Ψ meson and the η_c^- meson is much smaller than the mass difference between the ρ meson and the pion.

Let us investigate the baryons. Here, too, we have an opportunity to compare hadrons of different spins. For example, the only difference between a nucleon and the Δ particle is that the three quarks in the Δ particle are aligned whereas the two quarks in a nucleon always have opposite spin. We observe that the Δ particle is heavier than the nucleon.

No doubt this phenomenon must be related to the different spin structures of the two particles. Yet it turns out that the baryon situation is far more complicated than that of the mesons. In baryons, we cannot draw a parallel between electrodynamics and chromodynamics, which, of course, is not exactly news to us. There are no bound states of three electrons and there are bound states of three quarks. As we emphasized earlier, the reason for this is that the structure of the charges in QCD is more complicated than in QED because in QCD we have to contend

---

Figure 12.2

Mesons in which the spins of the quarks are aligned are heavier than mesons in which the spins are opposed. The reason for this is the chromomagnetic forces between quarks. Also, systems of three quarks with aligned spins are heavier than systems in which one of the three spins points in the opposite direction. This is why the Δ particle is heavier than the nucleon.

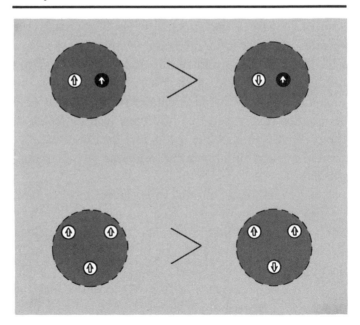

with eight different color charges. The chromomagnetic forces between quarks inside a nucleon or a Δ particle have to be added up just like the chromoelectric forces. If we do this, we come upon a surprising and very welcome result. Chromomagnetic forces repel each other if the spins of three quarks point in the same direction and attract each other if the spins oppose each other (figure 12.2). We emphasize that this effect is due to the quarks' color property. If the color charge did not exist, we would have the opposite effect. Thus the fact that the Δ particles are heavier than the nucleons can be explained by the colorful nature of quarks and their chromodynamic forces.

The fact that the Δ particle is heavier than the nucleon is very important for the structure of the world. If it were the other way around, the proton would decay into the $Δ^{++}$ particle, hydrogen would not exist, and the lightest element in the world would be the $Δ^{++}$ nucleus and two electrons around it. This world would differ substantially from ours.

XIII

The Fine Structure of Quarks

In physics we call an object elementary if there is no indication that it consists of even smaller units. However, this definition is relative. The various fields of physics have different conceptions of what is and what is not elementary. For example, it is irrelevant in many areas of physical chemistry whether atoms have specific internal structure or not; in this field, it suffices to consider atoms to be the smallest units of matter. Atomic physicists, on the other hand, are interested in the structure of atoms. For them an atom is not elementary but instead consists of the electronic cloud and the nucleus, and so the nucleus is therefore a basic particle. Nuclear physicists, now, are interested in the internal structure of the atomic nucleus and discover that it is not elementary at all, but consists of nucleons. Finally, elementary-particle physicists are ultimately interested in the structure of the nucleons; for them the nucleons consist of quarks.

The Fine Structure of Quarks

The question arises as to whether quarks and leptons are "real" elementary particles or whether they consist of yet smaller constituent particles. The answer to this question is still undecided. Experimental physicists have searched for a possible substructure of quarks and leptons, but so far without success. Nonetheless, it may well be that quarks and leptons do have common constituent particles. We shall return to this important question at the end of the book.

How Elementary Is the Electron?

Physicists have concentrated on the electron because it is the particle that can be most easily investigated in the laboratory. It seems to be consistent with our thinking to regard the electron as an elementary particle. No sign of an electronic substructure has been found to date. The electron seems to be a point-like object. If it has a finite radius, it must be less than 10^{-16} cm, that is, less than a thousand times smaller than the proton.

However, it turns out to be somewhat untrue to claim that an electron has no substructure whatsoever. As we have seen, an electron is surrounded by a vacuum polarization cloud, which in itself is a relatively complicated object. Therefore, an electron has something like a substructure as a result of its electromagnetic interaction. However, there is a big difference between this kind of substructure and, for example, that of a proton and its quark structure.

The substructure of the electron, it turns out, can be calculated with great precision simply by applying the laws of quantum electrodynamics. This calculable substructure is called fine structure. Many effects due to the fine structure of the electron have been studied since 1950. In most cases very satisfactory agreement between experiment and theory has been achieved. Of course, there have been instances where the two did not mesh, but in those cases it has turned out that the experiments were wrong or the theorists had miscalculated. Today the theory of quantum electrodynamics is the best theory we have in physics; its agreement with experimental data is impressive.

The Fine Structure of Quarks

What do we expect to be the case for quarks in QCD? According to the view of QCD, quarks are as elementary as electrons. They have no substructure, but, because of their interactions with gluons, they have a fine structure that is quite similar to that of electrons in QED. This means that quarks are not completely pointlike objects but are extended objects as a result of the chromodynamic interaction. This fine structure of quarks is as calculable as that of electrons.

One important test of the theory of chromodynamics is to measure the effects due to quark fine structure and to compare the results with theoretical predictions. In which experiments should we look for this fine structure effect, though? Clearly, it is much harder to observe a fine structure for quarks, which exist only inside hadrons, than to observe the fine structure of electrons, which exist as independent particles.

Recall the electron and neutrino scattering experiments described earlier, which proved that quarks act like pointlike, structureless particles inside the nucleon. If quarks indeed have a fine structure, it must be possible to see it in these scattering experiments. Also recall that quarks inside a nucleon moving at high speed carry only about 50 percent of the total nucleonic momentum. The remaining 50 percent must be accounted for by something else, something that has no electrical or weak properties. In QCD we have a candidate for the "something else"—the gluons, which are electrically neutral. Their only characteristic is that they carry energy and momentum. Thus we arrive at an important conclusion. About 50 percent of the nucleonic momentum is carried by the gluons, and therefore gluons must play an important role in the dynamics of the nucleon.

Suppose we scatter an electron off a quark inside a nucleon. In QCD the quark is not merely a quark, but a relatively complicated object with a fine structure. If the electron collides with a quark with relatively little momentum, the electron will not see the fine structure of the quark; that is, the electron will slide off the quark as though the quark were a pointlike object. However, if we increase the energy of the electron, we slowly begin to discern the fine structure of the quark. Specifically, the quark will occasionally look like a quark surrounded by gluons.

The reader may ask how an electron can distinguish a quark from a quark surrounded by gluons. The answer is that the electron has no problem making such a distinction because it reacts only with the

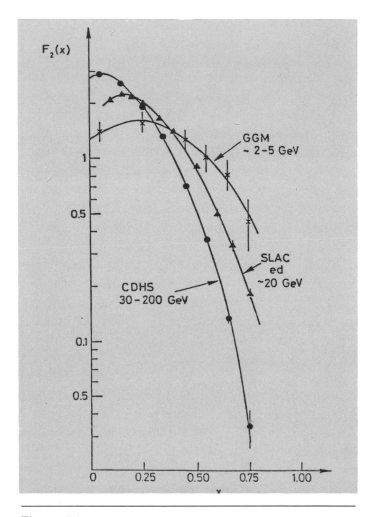

Figure 13.1

The distribution function of quarks in the nucleon as a function of the parameter x denotes the fraction of the nucleonic momentum (or nucleonic energy) carried by the quarks. As the reader can see, the function diminishes as the energy increases. This phenomenon is predicted by QCD theory.

The Fine Structure of Quarks

electrically charged quark and not with the gluons. The momentum of a quark surrounded by gluons will be less than that of a quark devoid of gluons.

By studying the results of the electron or neutrino scattering experiments, we can deduce how quarks are distributed inside the nucleon. Specifically, we can see the distribution of the momenta inside a quark. The fraction of the total nucleonic momentum carried by a quark is usually denoted by the parame-

---

Figure 13.2
A view of the CDHS detector at CERN, named after the collaboration of physicists from CERN, Dortmund, and Heidelberg (Germany), and Saclay (France). Using this huge detector, the physicists were able to study subtle details of the interaction of neutrinos and quarks and to verify the predictions of QCD.

ter x, which can vary between 0 and 1 (for example, $x = 1/2$ means that the corresponding quark carries half the nucleonic momentum). We just argued that the contribution of a quark to the total nucleonic momentum becomes smaller as the energies increase, which implies that we observe fewer and fewer quarks carrying a high momentum. Specifically this means that the function describing the distribution of the quark momenta decreases as the energy increases (figure 13.1).

Experiments are in excellent agreement with the theoretical calculations based on QCD theory (figure 13.2). Specifically, it has been possible to estimate the value of the QCD coupling constant α_s. It turns out that α_s for the electron and neutrino experiments is about 0.2, which is much larger than the electromagnetic fine structure constant $\alpha = 1/137 = 0.0073$. Of course, this is no surprise to us since the interaction of quarks inside a nucleon is stronger than the interaction of electrons inside an atom. Nevertheless, $\alpha_s = 0.2$ means that the quarks seen in the electron or neutrino scattering experiments behave largely as independent, nearly pointlike objects. The interaction of quarks with each other has become rather weak, and—as predicted by the chromodynamic theory of quarks and gluons—the strong interaction is about to vanish.

XIV

A Surprise at PETRA: Quarks Become "Visible"

The annihilation of electrons and positrons, which generates pure energy, is an especially significant process in high-energy physics because this energy can be used to produce many different particles whose total energy equals the total energy input. This annihilation process differs considerably from others. For example, proton-proton scattering always has two nucleons "left over" (as well as some other particles, mostly mesons). For this reason, proton-proton scattering is less effective than electron-positron annihilation in providing us with new knowledge about the structure of matter. With proton scattering, we only learn more about the proton itself. Everything that can be produced in proton-proton collisions, the proton already contains.

Investigating the Vacuum

This is not the case with electron-positron annihilation, however, where there is the possibility of producing exotic new particles from the available energy. It is accurate to say that these new objects are "created out of nothing," that is, out of the vacuum. For this reason, particle physicists sometimes say that electron-positron annihilation is a tool to investigate the structure of the vacuum. (We have already noted that the vacuum can be a rather complicated state for particle physicists.)

As an example of the production of new particles in electron-positron annihilation, we cited the production of charm. Given sufficient energy, we have no problem producing pairs of charmed particles in electron-positron annihilation. However, it is no easy matter to produce charmed particles in hadronic collisions. In fact, it was not until 1979 that physicists even discovered that charmed particles can be produced in such collisions (for example, proton-proton scattering). Because there are so many other particles produced along with the charmed particles, detection of the latter is rather difficult.

Not so in electron-positron annihilation, where we can produce a pair of charmed particles and nothing else. This is a very clean process and therefore a much more convenient one for the study of charmed particles.

A hadron—a ρ meson, for instance—can be produced in the electron-positron annihilation process in only one way: through the initial production of a quark-antiquark pair via the electromagnetic interaction. This is so in QCD because quarks are carriers of electric charge, and such a process will produce only elementary objects with electric charge. For example, there is no way of producing gluons in this process because gluons are neutral. Of course, when I say that we produce quarks in electron-positron annihilation, I do not mean that we really produce isolated quarks. What I mean is that we initially produce a quark-antiquark pair, that right away turns into one or several hadrons via the chromodynamic interaction between quarks. At first the quarks are so close to each other, however, that we may ignore the effects of chromodynamic interaction, at least if sufficient energy is available. The total rate for producing hadrons is then given by the quark charges, and we arrive at the prediction of color quark theory discussed in chapter 9.

The Ring at PETRA

In 1976 a committee of physicists, including myself, gathered at the DESY laboratory in Hamburg to discuss the research program for the new PETRA storage ring that was just being constructed at the DESY site. The plan was to have electrons and positrons collide with each other at energies of approximately 20 GeV each, which means that the total energy of the colliding particles would be twice that amount. Our discussion concentrated mainly on the question of what such a particle system looks like.

Suppose we take a detailed look at an annihilation process at such high energies. An electron and positron, both having the huge energy of 20 GeV, collide and annihilate each other, sometimes producing a quark-antiquark pair, a c quark and its antiquark, for example. Each quark will have the same amount of energy as the incoming electron and positron, namely, 20 GeV.

From what we already know, it is easy to imagine what happens next. We have already discussed the gluon string that holds quarks together and generates the large color forces between them. However, a long gluon string such as we discussed exists only in the case of the fictitious heavy quarks we used in our thought experiment. In the real world, such a string does not exist because the light quarks u and d do exist, and this changes the situation as follows. Suppose we separate two heavy quarks—a charmed quark and its antiquark, for instance—from each other. Initially the color force between them becomes increasingly stronger, and so the force of attraction remains constant as the separation distance increases. But this building up of the color string force costs energy, which has to be supplied through the annihilation of the electron and positron.

The physical vacuum is filled with an infinite number of virtual quark-antiquark pairs simply waiting to manifest themselves in the form of physical particles. This is feasible, however, only if the corresponding energy needed for this manifestation is available. It turns out that the virtual quarks and antiquarks in the vacuum are exceedingly selfish. They steal the energy we put into the system to separate the two initial quarks and to build up the color string. As a result, the initial quarks (or antiquarks) pick up one of the antiquarks (or quarks) from the vacuum and form one or several mesons. This means the end of the color string. Once the mesons, which are color singlets, are formed, there is no further need for the color field lines to connect the two initial quarks. Finally we observe nothing but a number of mesons.

For example, let us consider the production of a charmed quark and its antiquark in the electron-positron annihilation process. Both quarks have a large energy and fly away from each other at a speed close to the speed of light. In no time at all, a c quark will have picked up a $\bar{u}$ or $\bar{d}$ quark in the vacuum and formed a D meson. Likewise, the $\bar{c}$ quark will have picked up a u or d quark and formed another D meson. At the end, therefore, we simply observe the production of a pair of D mesons, together with other particles (perhaps several pions, figure 14.1).

Quark Jets

What does the final hadronic system look like? If the initial energy is large, on the order of 20 GeV, for example, it is possible to produce a large number of particles. Besides the D mesons, we can have many pions, K mesons, proton-antiproton pairs, and so forth. How are the momenta of these particles distributed? Do the particles that emerge from the region of interaction fly off in all directions or do they prefer certain specified directions?

The theory of chromodynamics allows us to predict the momentum distribution of particles created in annihilation experiments. What we find corresponds to what one guesses: the momenta of the hadrons at the

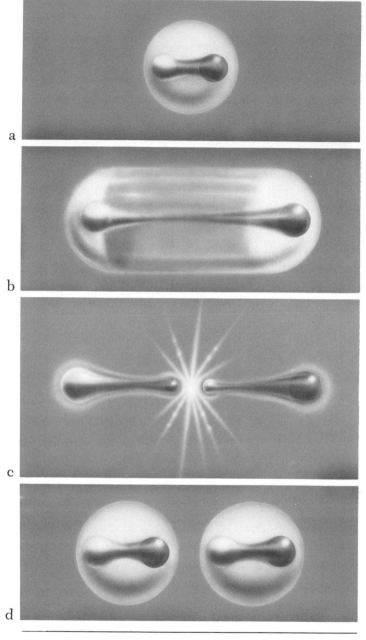

Figure 14.1

A c quark and its antiquark are produced in the e^+e^{--} annihilation at high energy (a). Both quarks move away from each other (b). As a result of the chromodynamic interaction a new $\bar{u}u$ or $\bar{d}d$ pair is produced (c). Those "light" quarks combine with the c quarks to form colorless hadrons, a pair of D mesons (d).

A Surprise at PETRA: Quarks Become "Visible"

end of such a process correlate with the momenta of the initially produced quarks. The particles are produced by what is called the fragmentation of the quarks. In other words, what we observe is two particle "jets," with the momentum of each jet equal to that of each of the initial quark momenta (figure 14.2). A particle jet is therefore nothing but a number of hadrons flying off in roughly the same direction. These jets may be interpreted as "decay products" of the quarks, and for this reason we often

---

Figure 14.2
An electron and a positron annihilate each other and produce a high-energy quark-antiquark pair. The latter fragments into hadrons (pions, K mesons, and such). The final hadronic system consists of two particle jets, which we call quark jets.

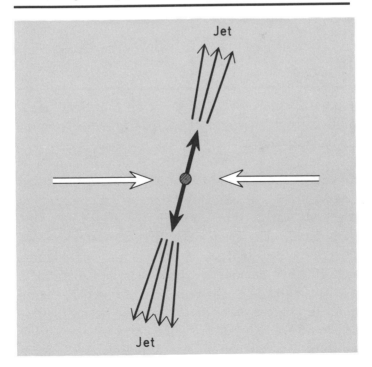

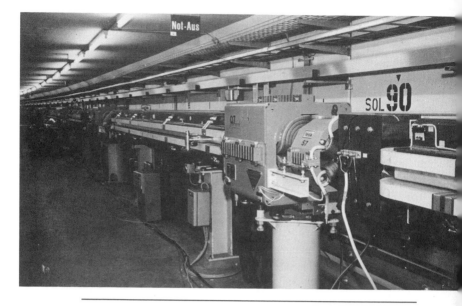

Figure 14.3
A look inside the PETRA tunnel, showing the vacuum pipe in which electrons and positrons can be accelerated to energies of up to 20 GeV.

speak of quark jets. Therefore, we can "see" quarks indirectly in the form of the quark jets of electron-positron annihilations.

The first hint that electron-positron annihilation produces quark jets came in 1975 at the SPEAR storage ring at SLAC. However, since the energy at SPEAR was not great enough, the evidence was not absolutely convincing. The first unmistakable evidence for the existence of jets was not obtained until the PETRA machine at DESY went into operation in 1978 (figures 14.3 and 14.4). It was observed that, at energies on the order of 15 GeV per beam, the final hadrons are distributed in the form of narrow jets. For the first time, physicists was able to "see" quarks,

Figure 14.4
The TASSO particle detector at the PETRA ring. Electrons and positrons annihilate each other inside the detector, producing electrons, muons, or hadrons.

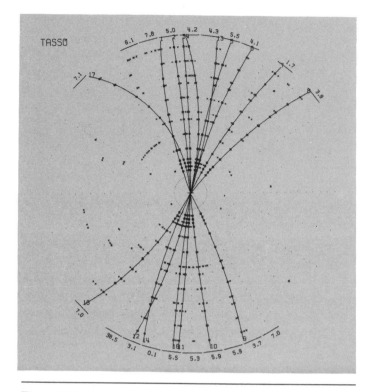

Figure 14.5

An event observed by the TASSO detector in 1979 during electron-positron annihilation at PETRA. The energies of the positron and electron were 17.5 GeV each. We can clearly see the two quark jets, which consist chiefly of pions (the numbers indicate the momentum of the particles). It is in these quark jets that quarks are indirectly "seen."

at least indirectly. Figure 14.5 shows an annihilation event observed at the PETRA-ring at DESY.

The results obtained by detectors at PETRA and other storage rings are fascinating and in perfect agreement with QCD theory. The quarks that manifest themselves indirectly in quark jets have become almost real. They are subjects of investigation despite the fact that they spend their lives locked inside hadrons.

XV

Smashing
the Proton

We have just discussed how we can see quarks indirectly if we scatter electrons or neutrinos off the quarks in protons. Still, it would be useful to obtain further evidence for the quark structure of the nucleon. How shall we go about doing so?

Let us again consider the scattering of electrons off nucleons, but this time in greater detail. Specifically, we shall assume that the electrons used for the experiments have much more energy than what is experimentally available anywhere at present. Specifically, let us suppose that the electrons we use have an energy of more than 100 GeV. If an electron of such high energy scatters off a nucleon, it does so because it has hit one of the quarks inside the nucleon and has transmitted to the quark a large amount of energy, several tens of GeV. In the experiments we have mentioned so far, for example, the SLAC experiments or the various neutrino or muon scattering experiments, physicists have looked for only the escaping lepton (an electron in the SLAC experiments and a muon in the neutrino and muon experiments). The energy and momentum of the final leptons were measured, and in this manner information was obtained about the quark distribution inside the nucleon.

More Quark Jets

We shall now do something else. Besides looking at the final lepton, we shall study in greater detail what happens to the nucleon after it has been hit by the incoming lepton. Before the scattering, the proton consists of three quarks more or less at rest with respect to one another. However, this situation changes radically once the proton has been hit by the electron. The electron transmits a large amount of energy to the quark it encounters, and we obtain the following quark configuration: two of the quarks remain essentially unchanged, but the third, the one that has absorbed the energy, moves away at high speed (essentially, the speed of light).

We recall encountering a similar situation in the electron-positron annihilation discussed previously. There we described what happens when a quark and antiquark escape each other at high speed: we get two quark jets. With that in mind, it is not difficult to imagine what happens in the electron-proton scattering experiment: we expect the "escaping" quark in this instance to again eject in the form of a jet (figure 15.1). The final hadronic system, the debris left from

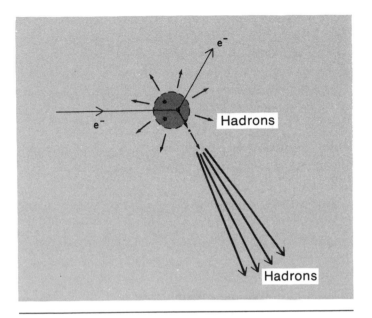

Figure 15.1
A high-energy electron hits a quark inside a proton. The quark is ejected from the proton (essentially at the speed of light) and fragments into hadrons, which form a quark jet.

the scattering of the proton, consists of the following particles:

1. The fragments of the quark hit by the incoming lepton
2. The relics of the two quarks not hit by the electron (these form particles whose momentum and energy we expect to be rather small and diffused in all directions).

The creation of quark jets in electron-proton scattering should enable us to see the quarks in the nucleon more directly than was possible in the SLAC experiments discussed in chapter 6. Therefore, scientists have been on the lookout at SLAC, FNAL, and CERN for jets in final hadronic systems. While what they have observed so far does not contradict theoretical expectations, it does not confirm them either. Our calculations demonstrate that, when leptons hit protons, more than 500 GeV of energy is needed to produce visible jets. Such energies are not yet available at high-energy laboratories. For example, the maximum energy for neutrinos at CERN is about 200 GeV.

The situation is expected to change in the future. At the end of the book, we shall discuss the HERA proposal, a plan to construct a high-energy accelerator that simultaneously accelerates protons and electrons. If our concepts about the quark structure of the nucleon are correct, this machine will essentially be an electron-quark collider. After the collision, the quark will be ejected from the nucleonic system and fragment into a quark jet. This quark jet should be highly visible in the HERA machine. Unfortunately, HERA will not be ready for experiments before 1992.

So far we have said little about protons scattering off nucleons, and the reason for that is simple. In studying the structure of the quark, we are better off using leptons than protons because leptons are elementary particles whereas protons are complicated three-quark systems. It is nearly impossible to describe in simple terms what happens if two protons collide, if only because generally all the quarks collide with each other. Still, under certain circumstances, which we shall discuss presently, it is possible to do so.

At present, the highest energies for proton-proton scattering experiments can be produced in the CERN

Smashing the Proton

Figure 15.2
The ISR accelerator at CERN. This machine can accelerate protons to an energy of nearly 30 GeV. Subsequently, the protons collide in special interaction regions. The ISR will discontinue its operation in 1984 (courtesy CERN).

Intersecting Storage Ring (ISR, figures 15.2 and 15.3). In this machine, protons of roughly 30 GeV collide head on.

Until recently, high-energy physicists believed that if two protons were scattered off each other at high energies, they would "heat up" and ultimately produce a variety of hadrons that would then scatter in the same directions as the incoming protons. Specifically, we expected the transverse momentum (the momentum of the hadrons transverse to the direction of

Figure 15.3
One of the interaction regions at ISR. The two intersecting vacuum tubes in which the protons move essentially at the speed of light are clearly visible. The protons collide at the point of intersection.

the incoming protons) to be very small, on the order of a few hundred MeV (figure 15.4).

Because of these expectations, then, the results of the first experiments at CERN-ISR in 1973 astounded everyone. In these experiments, hadrons with rather large transverse momenta were frequently produced (up to 5 GeV and more). How is this possible? The only explanation is that nucleons must contain hard points that do not break up at once. Conceivably, such hard points collide with each other and

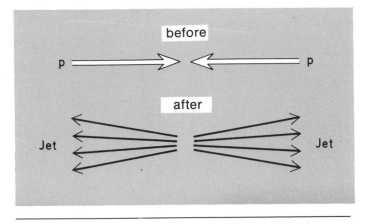

Figure 15.4
A schematic of proton-proton scattering at high energies. We can see two particle jets with momentum pointing in the direction of the incoming protons. The momentum of the particles transverse to the incoming protons is relatively small (typically less than 1 GeV).

produce large transverse momenta—the reader is invited to think of two billiard balls colliding.

But what are we to make of the hard pieces in the proton? It seems natural to identify them as quarks. Perhaps two quarks collide in a hard scattering process, and we find ourselves in a situation similar to the one we encountered in the description of lepton-nucleon scattering. The two colliding quarks are both knocked out of their protons, producing two quark jets that move with relatively large momenta transverse to the original direction of the proton (figure 15.5). If this picture is correct, the particles with large transverse momentum must all derive from the fragmentation of quarks that have collided violently. This means that particles carrying large tranverse momenta must belong to quark jets. Specifically,

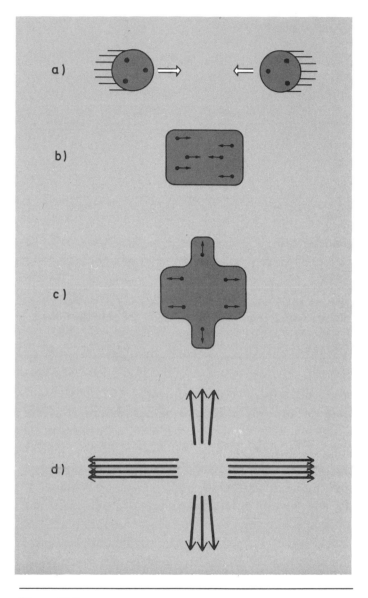

Figure 15.5

Four particle jets in proton-proton scattering. (1) Two protons accelerate toward each other. (2) The protons fuse and form a six-quark system. (3) Two quarks collide head on and are knocked out of the hadronic system. The other four quarks continue their flight undisturbed. (4) The final result is two particle jets created by the quarks that remained relatively undisturbed (left and right) plus two quark jets consisting of particles with large transverse momentum (up and down).

such particles will never be produced by themselves, but always as parts of a jet. Furthermore, the final hadronic state in such a case should always consist of four jets (figure 15.5).

Hard Quark-Quark Scattering

CERN-ISR and FNAL have conducted experiments in recent years to see if this picture corresponds to reality. In the meantime, it has been established that the particles that carry large transverse momenta are indeed never produced alone; they are always part of particle jets. Thus it appears that our theory of hard quark-quark scattering is correct. Many details still need to be explained, however, and we have to regard the present situation with caution. One particular problem is that the highest energy available for such experiments at CERN-ISR is still too low to identify the various jets. What we observe coincides with our picture of the hard quark-quark scattering, but we cannot yet say that the evidence speaks unequivocally in favor of this interpretation. We need further experiments at higher energies before we can be absolutely certain.

The Interquark Force

So far we have failed to mention what it is that causes quarks to scatter in a hard fashion. Naively we might assume that hard scattering is due to head-on collisions, as described in figure 15.5. In particle physics, however, matters are not that simple. Quarks cannot really collide with each other unless there is some force to make them interact in this fashion. The same is true for lepton-quark collisions, which we discussed above and which play an important role in the description of lepton-nucleon scattering.

An electron scatters off a quark inside a nucleon because it interacts electromagnetically with the quark. If there were no such thing as the electromagnetic force, the electron would pass right through the nucleon. Thus the question arises, What is this force between quarks that makes them collide so swiftly as to produce large transverse momenta in proton-proton collisions?

If we trust quantum chromodynamics, there is only one candidate for this force between quarks: the same force that is supposedly responsible for the binding of quarks into color singlet hadrons, namely, the gluonic force. We therefore have another possibility of testing chromodynamics by looking for quark-quark collisions in high-energy proton-proton scattering. In chromodynamics the properties of quark-quark scattering can be calculated with precision in terms of parameter as the QCD analog of the fine structure constant, which can also be measured in lepton-nucleon scattering. We mentioned earlier that the relevant QCD fine structure constant in the energy region under study in the accelerators is of the order of 0.2. We can use this information to calculate how often quarks collide in proton-proton scattering, and consequently we can determine how often we should be able to observe particles with high transverse momentum.

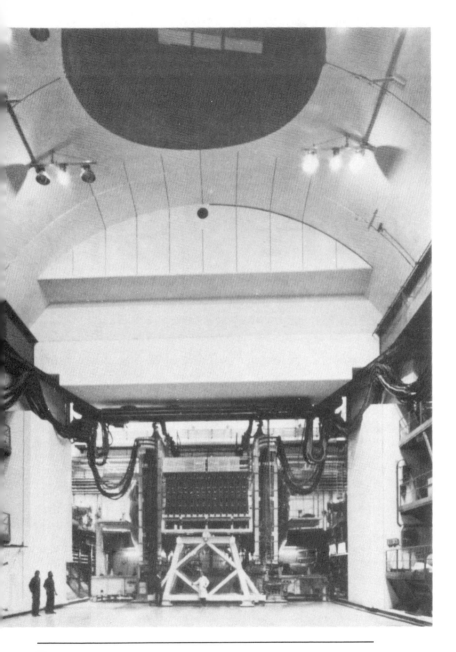

Figure 15.6
A view of the UA2-detector, one of the large particle detectors, installed more than 60m underground at CERN. The detector is placed in the SPS beam line. In its center protons and antiprotons collide head on (courtesy CERN).

Figure 15.7

A proton entering from the left collides with an antiproton entering from the right. Both the proton and the antiproton move essentially at the speed of light; they both have an energy of 270
GeV. This is one of the first events observed at the new CERN
collider. It is interesting to find out that many particles, mostly
pions, are produced in this reaction—one of the most violent collisions ever observed in nature (courtesy CERN).

Figure 15.8

An interesting event observed by the research group NA5 at the CERN collider in summer 1982. The event shows what happens after the collision of a proton and an antiproton both carrying an energy of 270 GeV. The particles leaving the interaction region nearly in the horizontal direction carry only a small transverse momenta. They are interpreted as the fragments of the incoming proton or antiproton. Besides those particles one observes a number of particles carrying large transverse momenta—they are flying off in vertical directions. These particles belong to jets produced by the fragmentation of quarks or gluons, which have undergone a hard scattering (courtesy CERN).

At CERN in 1981, a new accelerator ring was added to the existing 400-GeV accelerator called SPS. Using a newly developed ingenius technique, physicists can fill this ring with antiprotons that are subsequently accelerated to about 270 GeV and made to collide with protons of the same energy (figure 15.6). In 1982 the first preliminary results of these experiments were published. One of the first events observed at CERN is shown in figure 15.7.

For the first time, we can study the reactions of matter and antimatter at such high energies. In the summer of 1982 an interesting development took place at the CERN collider. Sometimes one observed the production of particle jets carrying very high transverse momentum (see figure 15.8). The experimental findings are in agreement with the predictions from QCD. A detailed quantitative check has not been made, however. Nevertheless, one can say that for the first time the hard collisions within the proton have been observed. Certainly the study of the collisions of protons and antiprotons at high energies will become an interesting new tool with which to investigate the physics of quarks and gluons in the future.

XVI

How to "See" Gluons

Over the last ten years physicists have obtained considerable evidence that quarks are the constituent particles of hadrons. In chromodynamics, however, we have not only quarks but also gluons. This being the case, the reader may ask if there is any way of seeing gluons—at least indirectly, the way we have been able to "see" quarks. Can we also detect gluons in proton-proton scattering experiments? This is an important question, of course. Everyone agrees that the evidence for the existence of gluons is slim compared with that for the existence of quarks. Gluons have remained quite inconspicuous so far.

One way of looking for gluons is to study the force between quarks, which is supposedly generated by gluons. We emphasized in previous chapters that the gluonic force in QCD is precisely what is needed to explain the many puzzles of hadronic physics, such as the binding of quarks into color singlets, for example. Phenomena such as this all represent indirect evidence that gluons exist.

If chromodynamics is the correct theory of the strong interaction, then there must be particles that consist of nothing but gluons. In the slang of high-energy physics, these peculiar objects are called glue balls, and, if they exist, glue balls must be color singlet hadrons without quark content. They must consist solely of glue, and the simplest color singlet configuration must consist of two gluons because, of course, a single gluon cannot form a color singlet system. (In QCD a gluon belongs to a color octet because the color force is generated by the exchange of eight gluons.) Three, four, or more gluons can be used to construct more complicated objects.

Since gluons lack electrical or weak properties and therefore do not participate in the electromagnetic or weak interactions, glue balls are necessarily rather inconspicuous particles. Specifically, they lack electric charge. Physicists have looked for glue balls in various proton-proton scattering and electron-positron annihilation experiments, but progress was made only very slowly.

In 1980, while investigating the hadronic decay of the J/Ψ particle, a group of physicists at SLAC headed by Robert Hofstadter of Stanford University found a new particle having mass of 1.4 GeV. This new particle has a number of properties that are consistent with our ideas about glue balls. However, we cannot be certain that this particle is not simply an ordinary meson consisting of a quark and an antiquark. It will probably take several years to settle the issue. The new particle has been named iota (denoted by the Greek symbol ι).

Gluons in J/Ψ Decays

As we have emphasized, the strongest evidence that matter is composed of quarks derives from our observation of quark jets in high-energy electron-positron annihilation processes. The jets are created because this process produces a quark and an ántiquark of high energy that move nearly at the speed of light. The question arises, Can we arrange something similar for gluons? How can we produce a gluon of such high energy and momentum that it will behave something like a quark and produce a jet of hadrons. Any jet thus produced would be a gluon jet and would carry the momentum and energy of the parent gluon.

One possibility of tracking gluon jets is to study the decay of heavy quark-antiquark objects like the J/Ψ particle. In chapter 8 we mentioned that the J/Ψ meson frequently decays into a pair of muons or into an electron-positron pair. These processes are manifestations of the electromagnetic interaction and are well understood. However, the J/Ψ meson can also decay via the strong interaction, for example, into pions. Let us concentrate on this decay produced by the strong interaction, and speculate as to how it must proceed according to QCD theory.

The J/Ψ meson consists of the charmed quark and its antiquark. How is it possible for this system to decay via the strong interaction into pions or other particles that consist not of charmed quarks but of the normal, "light" u, d, and s quarks? We have already mentioned that the dynamics of the J/Ψ meson are similar to those of positronium, the bound state of an electron and a positron. For this reason, we find it useful to take a somewhat closer look at the decay of positronium.

First of all, recall that there are two kinds of positronium: orthopositronium, with aligned particle spins, and parapositronium, with opposed particle spins. It is a peculiarity of the electromagnetic interaction that parapositronium and orthopositronium decay into two and three photons, respectively. Orthopositronium is prohibited from decaying into two photons as a consequence of Maxwell's equations, as discussed in chapter 2.

How then should a heavy quark-antiquark system like the J/Ψ meson (also sometimes called charmonium) decay? According to QCD theory, this decay should resemble that of the three-photon decay of positronium: it should produce three gluons. Of course, this does not imply that we shall literally see three gluons in the decay of this heavy quark-antiquark system. Since gluons also have color characteristics, they cannot really be produced. What we mean is this. In the decay, gluon quanta will be produced first. After these quanta get a certain distance from each other, the confining (infrared slavery) forces will begin to take effect and the gluons will fragment into hadrons. The final hadronic system will consist of three hadron jets, each jet being the debris of the original gluon (figure 16.1).

In a J/Ψ meson, the energy of the gluons averages out at 1 GeV, which is much too low for us to observe

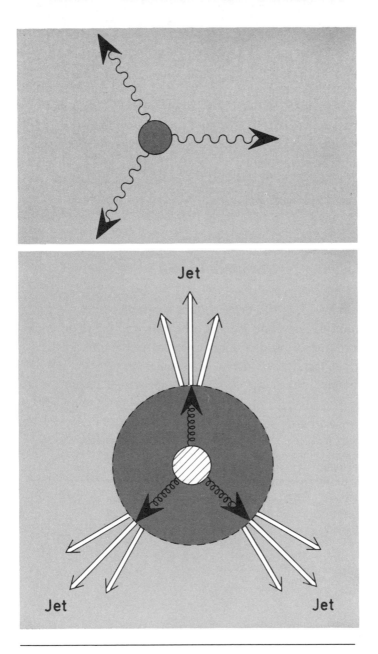

Figure 16.1
The decay of orthopositronium into three photons and the corresponding decay of charmonium (J/Ψ meson) into three gluons. Since gluons are confined objects carrying color, they fragment into hadrons and form three gluon jets.

gluon jets. The only way we shall eventually be able to observe such jets is studying the decay of quark-antiquark resonances which are heavier than the J/Ψ, provided such systems exist in nature. Of course, such systems exist only if there are quarks with an effective mass larger than that of the charmed quark. Are there such quarks?

Looking for Another Quark

In 1977, a 400-GeV proton beam was directed at a nuclear target at FNAL, the purpose being to look for pairs of muons. This experiment, conducted by a group headed by Leon Lederman (figure 16.2), resembled the Ting experiment described earlier, the

Figure 16.2
Leon Lederman, the leader of the group that discovered the Υ meson. Lederman is currently director of the Fermi National Accelerator Laboratory (Fermilab or FNAL) near Chicago.

essential difference being that Ting had studied the production of electron-positron pairs and the group directed by Lederman was to concentrate on muon pairs.

The Lederman group came upon an astonishing result: far more muon pairs of one particular energy were produced than anticipated. The muon pairs peaked in their masses at one particular energy: 9.5 GeV. The only explanation for this phenomenon is that there is produced at this energy a particle that is very similar to the J/Ψ meson. In other words, the group had found a new meson, one with a mass of 9.5 GeV, ten times heavier than the proton. The new particle, called the upsilon meson (hereafter referred to as the Υ meson), could be a bound system consisting of the new quark and its antiquarks. Of course, this was just a hypothesis and so had to be tested. One way of testing whether a new quark had been discovered was to try to produce it in the electron-positron annihilation process. If indeed it could be produced and had properties similar to the J/Ψ meson, then there could be no further doubt.

In early 1978 a special program was started at DESY to reach the required energy of 9.5 GeV. The DORIS storage ring at DESY, originally designed to produce a maximum energy of 4 GeV per beam, had to be modified to increase the energy to 5 GeV per beam. In June 1978 the experimenters approached the Υ energy region, and the Υ meson was observed at an energy of 9.46 GeV (one of the particle detectors which played a significant role in this discovery is shown in figure 16.3). As expected, it behaved much like the J/Ψ meson, that is, it is a bound state consisting of a new quark and its antiquark.

Figure 16.3
A view into the interior of the Pluto detector at PETRA in Hamburg.

How to "See" Gluons

Careful studies of the ϒ meson in electron-positron annihilation indicated that the electric charge of the new quark is not 2/3, as that of the charmed quark is, but −1/3, like that of the strange quark. The new quark is called the b quark, and the ϒ meson is interpreted as a b̄b state.

While elementary-particle theorists had strong reasons to suspect the existence of the charmed quark, no such expectations had been entertained for the b quark. It has entered physics out of the blue, as it were. In the absence of theoretical arguments, therefore, we have to await further experimental findings before we can say what its properties are (for example, what the nature of its weak interaction is). We shall discuss what physicists think about the b quark in a subsequent chapter.

Three Jets in the Decay of ϒ

The ϒ meson is approximately three times as heavy as the J/Ψ meson. In its strong interaction decay, the three gluons share about one-third of the total energy of the ϒ system, that is, a little more than 3 GeV. But even this amount of energy is not great enough to allow us to discover the three jet structure we are searching for. However, refined analysis of ϒ meson decay allows us to detect the existence of the jets, at least indirectly. We know that electron-positron annihilation at high energies initially produces the two-jet structure already observed so clearly. If we tune the energy so that we are producing the ϒ state, however, we should no longer get a two-jet structure because

215

the decaying ϒ meson leads to three gluons emitted in more or less all directions. This is precisely what was observed at DESY: the two-jet structure vanishes as we come upon the ϒ energy. It appears that the decay of the ϒ state must initially proceed via the production of more than two quanta.

Another test of the three-gluon hypothesis goes as follows. The three gluons produced in ϒ decay are not emitted arbitrarily in three directions. All three must lie in the same plane because the three momentum vectors of the three gluons must add up to zero. After all, the original momentum of the ϒ state is zero (recall that three vectors whose sum is zero must lie in one plane). Therefore we expect the final hadronic state of ϒ decay to have a disc-shaped structure (figure 16.4). The momentum vectors of all final hadrons

---

Figure 16.4

The decay of the ϒ meson as a result of the annihilation of an electron and a positron. The momenta of the final hadrons form a disc-shaped structure, which is what we expect in QCD when the final hadronic system is produced via the annihilation of $\overline{b}b$ into three gluons.

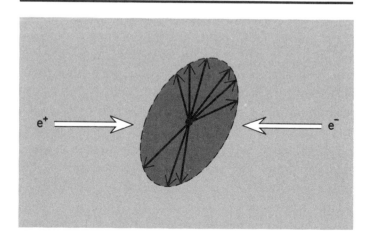

should form something approximating a plane. Of course, this plane changes from event to event, but each event should have only one plane.

The DESY physicists have taken a detailed look at the final state and found that the particle momenta do indeed describe something like a disc. This is a very welcome result since it corroborates the three-gluon emission hypothesis. However, the very best way to test these ideas would be to find another heavy quark-antiquark system, one even heavier than the Υ meson. If we can find such a state, at about 50 GeV, perhaps, we shall have an excellent chance of observing the three-jet structure caused by three-gluon decay. Unfortunately, no new heavy quark-antiquark system has been found thus far (fall 1982). At the PETRA ring scientists have searched for heavy quark-antiquark systems in the energy region from 12 to 37 GeV without any success. If such particles exist in nature, they must be heavier than 37 GeV.

Bremsstrahlung of Gluons

Let us consider another possibility of observing gluons indirectly. Electron-positron annihilation at high energy initially produces a quark-antiquark pair that can be regarded as a pair of free (noninteracting) quarks. Nonetheless, one of the quarks may "remember" in this process, that it has the option of interacting with a gluon. It may suddenly emit a gluon, which will carry a certain part of its energy and, as a result, decelerate. Such a process is called *bremsstrahlung,* and it is a well-known process in electrody-

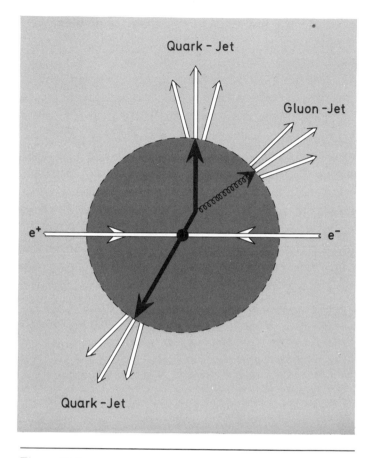

Figure 16.5

The production of a gluon jet in the electron-positron annihilation process. The lepton pair annihilates and produces a quark-antiquark pair. One of the quarks emits a gluon and therefore loses part of its energy and momentum and also changes its direction. The emitted gluon fragments into hadrons, which form a gluon jet. The final hadronic system consists of three jets, one of which is the gluon jet.

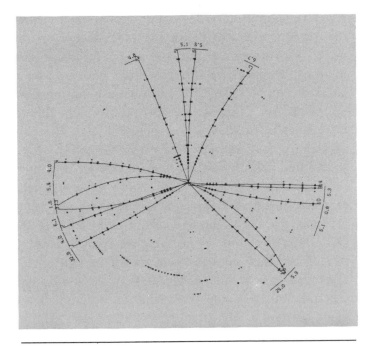

Figure 16.6
A three-jet event in electron-positron annihilation at PETRA. The energy was about 15 GeV per beam. The three jets are clearly visible. This event is interpreted by QCD as the result of the initial production of a quark-antiquark pair. One of the quarks emits a gluon in a bremsstrahlung process.

namics. Electromagnetic bremsstrahlung processes have been studied in detail both theoretically and experimentally. The agreement between theory and experiment is remarkable.

One example of electrodynamic bremsstrahlung is the emission of a photon from a muon subsequent to the production of a pair of muons in an electron-positron annihilation. Consequently, the momentum and flight direction of the muon that emits the photon are changed. We expect an analogous situation in QCD.

A quark may emit an energetic gluon shortly after the quark is created (figure 16.5). Subsequently, the quark, the antiquark, and the gluon fragment into hadrons. Finally, we observe three hadron jets: a quark jet, an antiquark jet, and a jet originating from the gluon. We can estimate how often we should observe such three-jet events at high energies. Roughly speaking, about 10 percent of all annihiliation events should manifest three-jet instead of two-jet structure.

In the summer of 1979, the PETRA storage ring provided the first evidence that the three-jet events predicted by QCD theory exist (figure 16.6). If the QCD interpretation of the three-jet event is correct, two of the three particle jets shown in figure 16.6 must derive from the fragmentation of quarks, and the third must be the fragmentation product of the gluon emitted in the bremsstrahlung process. Unfortunately, we cannot tell by looking at figure 16.6 which jet was produced by a quark and which by a gluon. In principle, it is possible to separate the quark jets from the gluon jets, but to do so we need to investigate the properties of the various jets (charge and particle distribution and so forth), and this has not been accomplished as yet.

Let us summarize the situation. If QCD is the correct theory of hadrons, it must be possible to observe gluons indirectly by looking for gluon jets. The best evidence for the existence of gluons derives from the electron-positron annihilation process, where we have observed three-jet events at high energies. These events are interpreted as manifestations of the bremsstrahlung process. The observation of the three-jet events is a welcome confirmation of the chromodynamic theory of hadrons.

XVII

Weak Interactions of Leptons and Quarks

Our discussion of quarks would be incomplete without the long-promised look at the weak interaction. Quantum chromodynamics, our theory of the strong interaction, is an example of a non-Abelian gauge theory. This theory uses the relevant gauge group SU(3), under which all color transformations can be subsumed.

It turns out that the modern theory of the weak interaction is also a non-Abelian gauge theory, albeit of a different type than QCD. As a matter of fact, some of the most exciting developments in particle physics in recent times have occurred in the field of weak-interaction physics. We shall examine some of these developments in this chapter.

First of all we need to summarize the main features of weak interactions as we have observed them experimentally. Right now we shall focus our attention on the u and d quarks and on the electron and its neutrino. All interactions involving these particles fall into one of two groups:

1. Processes that alter the electric charge (for example, β decay: n $\rightarrow$ p + e$^-$ + $\bar{\nu}_e$).
2. Processes in which there is no change in electric charge (for example, neutrino scattering ν_e + p $\rightarrow$ ν_e + p).

Type 1 processes are called charged-current reactions, and those of type 2 are called neutral-current reactions. In all these processes, precisely four fermions participate in the interaction. For example, β decay has one incoming fermion, the neutron, and three outgoing fermions, the proton, the electron, and the electron-antineutrino. In the type 2 example given above, we have two incoming and two outgoing fermions.

Another important feature of weak interactions is that they all occur with the same strength. The constant that describes the strength of the weak interaction is called the Fermi constant. Unlike the fine structure constant of electrodynamics, which is a pure number, the Fermi constant is a parameter with the dimension (energy)$^{-2}$. The empirical value of the Fermi constant is $G = 1.16 \times 10^{-5}$ GeV^{-2}. This is quite a small number (if we describe it in GeV units), which implies that weak interactions are indeed very weak, much weaker than electromagnetic interactions, at least in reactions where energies of the order of a few GeV or less are relevant.

Problems at High Energy

The fact that the Fermi constant is not a pure number is rather unsatisfactory. It implies that we shall not be able to base a consistent theory of the weak interaction on the interactions of four fermions described above—at least not a theory that accounts for weak interactions at high energies. For electrodynamic interactions, we have QED, which allows us to make predictions even for energies of 10^7 GeV or more. This is not the case for weak interactions. The simple picture of the interaction of four fermions we described above begins to come apart once we consider the weak processes at energies that correspond to the energy scale expressed by the Fermi constant, namely, the square root of its inverse, which is about 300 GeV.* This implies that at energies greater than about 300 GeV our picture of the weak interaction will no longer apply. It must be replaced by something else, but by what?

One possible solution to this problem—and, as we shall see, probably the correct one—is based on the following considerations. Let us compare the electromagnetic and the weak interaction. The basic electromagnetic interaction is that between photons and charged particles. The interaction of fermions—the scattering of two electrons, for instance—is described by the exchange of a virtual photon, which mediates the interaction between the two electrons. The idea is to describe the weak interaction between fermions in a similar manner through the introduction of two par-

---

*Since $G = 1.16 \times 10^{-5}$ GeV^{-2}, the square root of its inverse is $\sqrt{1/1.16 \times 10^{-5}} = \sqrt{86\,200} = 294$ GeV ≈ 300 GeV.

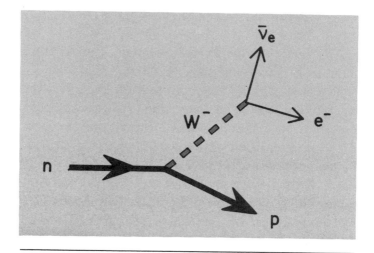

Figure 17.1
A possible description of neutron decay. The incoming neutron emits a virtual W⁻ boson and turns into a proton. The W⁻ boson disintegrates immediately into an electron and an antineutrino.

ticles that will mediate the interaction. For example, we may describe β decay in the following way (figure 17.1). The neutron changes into a proton and emits a virtual charged particle, called W⁻, which subsequently disintegrates into an electron and an antineutrino. This process is very similar to the electromagnetic interaction between two charged particles described in figure 2.2. Instead of the virtual photon, however, it is a virtual W⁻ particle that mediates the interaction.

The analogy between electromagnetic and weak forces becomes even clearer if we describe the neutral-current interactions by the same kind of analogy. To do so, however, we need a new neutral particle to mediate the interaction: the Z boson (figure 17.2). Indeed, life would be much simpler if the W−Z

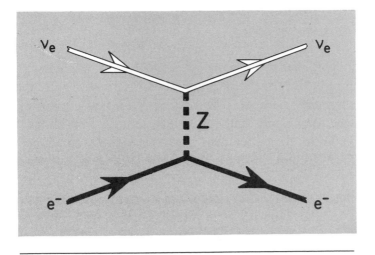

Figure 17.2
The neutral current process mediated by a virtual Z boson. The incoming electron-neutrino emits a virtual Z boson, which reacts with the electron.

scheme we just proposed really did describe the weak interaction.

The question arises as to whether the W and Z particles really exist. During the past fifteen years, experimentalists have studied many properties of the weak interaction, especially of neutrino interactions. We can set some limit on the properties of these particles. For one thing, neither the W nor the Z particle can be massless. As a matter of fact, experiments to date indicate that their masses are quite large: both have to be heavier than 50 GeV, that is, they have to be heavier than about fifty-five protons. Recently the W particle was discovered at CERN (see p. 245).

The Unification of the Electromagnetic and Weak Forces

The idea of introducing the W and Z particles as mediators of the weak interaction is part of the dream physicists have of constructing a unified theory of electromagnetic and weak interactions. We already know of many features that these two interactions have in common. Both are mediated by spin 1 objects (the W and Z particles must possess spin 1 to reproduce the observed features of the weak interactions), and both act with a universal strength, described by the fine structure constant α in electromagnetism and the Fermi constant G in the weak interaction.

The idea of unifying the electromagnetic interaction with other forms of interactions is not new. One of the first physicists to note that the electromagnetic and weak interactions may be related was Enrico Fermi, who formulated the first theory of weak interactions in the 1930s. The idea of using massive particles in describing the weak forces goes back to the Swedish physicist Oscar Klein, who worked on these problems in the late 1930s.

The reader may ask how the electromagnetic and weak interactions can be united if the latter is so insignificant compared with the former? How can they be compared if the strength of the electromagnetic interaction is given by the constant $\alpha \approx 1/137$ and that of the weak interaction by G, a number with the physical dimension GeV^{-2}? It is useless to compare α with G. We have to compare α with the strength of the fermion-W or fermion-Z interaction, and this strength is provided not by G but by an expression that involves the mass of the W or Z particle: the heavier the W and Z particles, the weaker the inter-action. This is easy to understand. If we made the W and Z particles infinitely heavy, they would drop out of the world; that is, the weak interaction would cease to exist.

This is the crucial point. Perhaps the observed weakness of the weak interaction is simply a conse-quence of the mass of the intermediate W and Z par-ticles. Perhaps those masses are very large, in which case the basic strength of the weak interaction is not really small at all. It may even be on the order of α.

To test this notion, let us suppose that the interac-tion between W and Z particles and fermions is as strong as the electromagnetic interaction. We calcu-late the Fermi constant in terms of the mass of the W particle and find a mass of 37 GeV, which is equal to the mass of about forty protons.*

Speculations of this kind were first made at the end of the 1950s, but not many people took them serious-ly. A W particle with a mass of nearly 40 GeV seemed so enormous that it boggled the imagination. In any event, only a tiny minority of physicists dared to think that some day they might actually be able to test this idea experimentally by observing the W par-ticle directly.

Since then, however, the situation has changed drastically. We are now quite confident that the weak interaction and electromagnetism can be regarded as different manifestations of one and the same force. The details of this theory have been developed by the theorists, and we now have a framework that is as well understood as the theory of quantum electrody-

* The exact formula is: $G = \dfrac{\pi}{\sqrt{2}} \cdot \dfrac{\alpha}{M_w^2}$

namics. We are able to calculate all processes, including the weak processes, at very high energies. Since the electromagnetic and weak interactions affect the flavor of leptons and quarks, we refer to the combined theory of the weak and electromagnetic interactions as flavor dynamics. (It is also called the theory of flavor interaction or electroweak theory.)

A Theory of the Weak Forces

The modern theory of flavor dynamics cannot be presented in detail here; to do so would entail the introduction of too many mathematical formulas. However, I can at least sketch out the main ideas.

First of all, let us recall one of the basic properties of observed weak interactions—the universality of their strength. All weak processes can be described by a universal strength constant, the Fermi constant. This reminds us of a similar situation in electrodynamics. The electrodynamic interaction of charged particles is described by a universal strength because the charges of the particles are quantized in units of the basic electric charge e. (Numerically, this basic electric charge is equal to the charge of the positron.) Therefore, we introduce a weak-interaction analog of the electric charge, which we call the weak charge. The electric charge is described by the strength with which a photon interacts with a charged particle. The weak charge determines the interaction of a particle with the W or Z particle.

The role of the charge associated with the W particle may be understood in the following way. An electron, for example, can turn into a neutrino by emit-

Weak Interactions of Leptons and Quarks

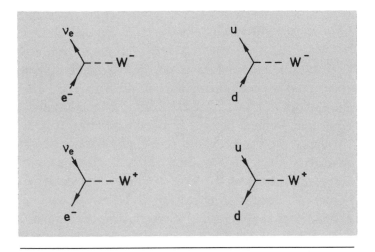

Figure 17.3

The elementary interactions of W bosons with quarks and leptons. An electron turns into a neutrino by emitting a W⁻ boson. Analogously, a d quark turns into a u quark by emitting a W⁻ particle. The W⁺ boson mediates in the conversion of a neutrino into an electron and in the conversion of a u quark into a d quark.

ting a virtual W⁻ particle. The electron-neutrino can emit a virtual W⁺ particle and turn into an electron. The weak charge associated with the W particles turns electrons into neutrinos and vice versa. The same holds true for quarks. The u quark can turn into a d quark by emitting a virtual W⁺, and the d quark can turn into a u quark by emitting a virtual W⁻ (figure 17.3). It is useful to think of quarks and leptons as the following twofold units:

$$\binom{\nu_e}{e^-} \qquad \binom{u}{d}$$

The weak charges associated with the W particles simply change the upper component of such a unit into the lower one, and vice versa.

The Problem of Parity

However, this picture is not entirely free of problems, and these problems have to do with parity. It is an important property of the electromagnetic interaction (one we have not yet stressed explicitly) that it is invariant with respect to a reflection of space, the so-called parity transformation. This simply means the following. Suppose we look at some electromagnetic process—the scattering of two electrons, for example. Now let us suppose we observe this process in a mirror. What we see looking in the mirror is the same process, of course. There is a question, however, as to whether the mirror-image process can actually occur in the real world. To find the answer, we must carry out another experiment, one that is a mirror image of the first. If we get the same result as what we saw in the mirror during the first experiment, we can say that the physical laws governing the process are invariant under space reflection or, in other words, that they exhibit parity invariance.

It has been known for a long time that mechanical processes are invariant under parity transformation (figure 17.4). The same holds true for electromagnetic interactions. One might think, as many physicists did in the past, that the invariance of the laws of nature with respect to parity is self-evident. However, it is not. Nature abounds with cases of left-right asymmetry. For example, the human body has the heart on the left side, and only very, very rarely on the right. If we look at a person through a mirror, the heart is on the right. Of course, we know that this is not the case in reality, but only a mirror effect. Nonetheless, we

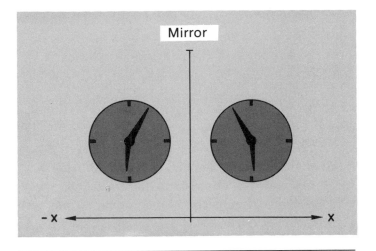

Figure 17.4
A parity transformation corresponds to the inversion of one of the space coordinates (in this case, the x axis). The clock on the left is the mirror image of the one on the right. The laws of classical mechanics are invariant with respect to parity transformation. For this reason, the motion of the two clocks is identical.

can imagine a human body with the heart on the right side, and this "mirror body" would not differ substantially from the real human body we are familiar with. All reactions of a mechanical, electrical, or chemical nature that occur within a "mirror body" lead to the same results as those in the original body.

No Parity for the Weak Force

At the beginning of the 1950s, some physicists began to inquire whether the weak interaction violates parity invariance or not. Two theorists, T. D. Lee and C. N. Yang, proposed several experiments to test the parity property of the weak force. By 1956 the answer was clear: the weak interaction violates parity.

After years of experimental work, it turns out that the parity violation observed in the weak interaction is rather simple. Let us consider a normal fermion, an electron, for example. We can decompose such a fermion in a left-handed and a right-handed part (figure 17.5). A left-handed fermion rotates like a left-handed screw; that is, its spin is directly opposed to its direction of motion. The spin of a right-handed fermion points in the direction of the motion of the elec-

Figure 17.5

The spin of a right-handed electron, called e_R, points in the same direction as the momentum of the electron, and the electron rotates like a right-handed screw. A left-handed electron's spin is directly opposed to the electron's momentum, and this electron rotates like a left-handed screw. Parity transforms a left-handed electron into a right-handed one and vice versa (as indicated by the double-headed arrow).

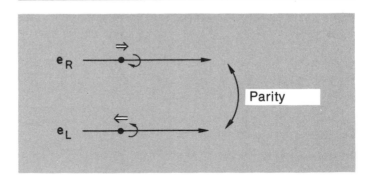

tron. Since in parity transformation a left-handed screw becomes a right-handed one, a left-handed fermion turns into a right-handed one. Therefore, it has become very useful to work with left-handed and right-handed fermions when we study the parity violation of the weak interaction.

We can try the same thing with the electromagnetic interaction. Since this interaction does not violate parity, however, the photon must interact equally with left-handed and right-handed electrons. (If it did not, we could use the interaction of photons with electrons to distinguish between left-handed and right-handed electrons.) A photon expresses equal preference for right-handed and left-handed electrons. The electromagnetic interaction is thoroughly democratic.

The weak interactions, however, drastically violate this democratic state of affairs. Among fermions, only the left-handed ones participate in the weak charged-current interactions. (We shall come to the special case of the neutral-current interaction shortly). The W bosons couple only with left-handed leptons and quarks. Right-handed leptons and quarks are entirely ignored. We have no idea why this is so and simply have to accept the left-handed nature of weak forces as an experimental fact.

We must also mention that the left-handedness of weak interactions holds not only for the electron and the neutrino and for the u and d quarks but also for the muon and its neutrino and for the s and c quarks.

XVIII

The Unified Theory
of Electroweak
Processes

We mentioned previously that a W boson can transform an electron into a neutrino and vice versa, which reminds us of a similar situation in chromodynamics. There we noticed that a gluon interacting with a quark can change the color of the quark. A red quark can turn into a green one, for example, because gluons have the same color properties as the eight charges of the color symmetry group SU(3). Let us assume something similar for the W bosons, which are supposed to be analogs of the gluons. We can describe weak interactions by a gauge theory, but to do so we have to find the symmetry group (gauge group) of the weak interactions we are dealing with.

The Unified Theory of Electroweak Processes

We determined the color group for the strong interaction simply by looking at the number of colors. Since there are three colors, the corresponding group is SU(3). We can determine the weak-interaction group by similar procedure. Since quarks and leptons always appear in two-member units, which we can call doublets, the relevant gauge group must include all transformations involving a doublet of leptons or quarks. This group is easily determined: it is the group SU(2). This group SU(2) also applies to the isospin, which is the approximate symmetry of the strong interaction (discussed earlier) that is so very important in nuclear physics. For this reason, we shall sometimes refer to the symmetry group of weak interactions as weak isospin. We must emphasize, however, that weak isospin is in no way related to normal isospin.

The group SU(2) has three different charges,* and two of them are easily identifiable with the weak charges that describe the interaction of the W^+ and W^- particles; that is, these two charges carry (like the W particles) an electric charge. The third SU(2) charge, however, is electrically neutral. How are we to interpret this third weak charge? Or, more generally, how can we describe weak interactions using the group SU(2) as their symmetry group? The proposition is a very simple one. We proceed just as we did with the strong interactions; that is, we interpret the weak interaction symmetry group as a gauge group and construct a gauge theory of weak interactions.

---

*Recall that in color group SU(3) we dealt with eight ($3 \times 3 - 1 = 8$) different charges. In the case of SU(2) we have three ($2 \times 2 - 1 = 3$) different charges.

A Miraculous Mass Generation

Weak interactions have one problem, however, which we have failed to face so far: the W particles, which will eventually be part of the gauge theory, cannot be massless. In general, as a matter of fact, the mass of the W particles is rather large. We did not encounter this problem with strong interactions because gluons, like photons, are supposed to be without mass. So the question arises as to how we are going to go about constructing a theory in which the corresponding particles (the W and Z particles) are massive. This problem can be solved by a trick: let us introduce yet another particle in the process, this time a scalar particle (that is, a particle without spin) to interact with the W and Z particles. We can arrange its interaction so that we are able to introduce a mass for the W and Z particles. At the same time, the weak-interaction symmetry is partly destroyed. For this reason, this procedure is referred to as spontaneous symmetry breaking. We shall not consider this mechanism in greater detail, but simply note that it can be used to introduce the masses of the W and Z particles.

The mechanism of spontaneous symmetry breaking gives us a consistent way of introducing the masses of the W and Z particles in our scheme. This was demonstrated in 1971 by Gerard 't Hooft, whose work (partly together with Martinus Veltman) has contributed significantly to the development of the theory of electromagnetic and weak interactions.

The simplest way of constructing a gauge theory of the weak interaction is to use the group SU(2) mentioned above, that is, to interpret the left-handed quarks and leptons as the doublets of the weak isospin. When we do so, we have three gauge particles, the two charged W bosons and a neutral particle that we identify as the Z particle. Note that the electromagnetic interaction is excluded from the scheme for a moment. We have constructed a gauge group exclusively for the weak interaction.

To find out whether this theory works, we have to consider the various experimental results. The theory makes a very simple prediction: since the charged W bosons interact only with left-handed quarks and leptons, we conclude that the same must be true of the neutral Z particle. In other words, we predict that neutral-current interactions affect only left-handed quarks and leptons.

What do the experiments show? For some time, it appeared as if neutral-current interactions, like charged-current interactions, prefer left-handed fermions exclusively. However, data that became available in 1977 demonstrated that this is not entirely the case: neutral-current interactions, though predominantly left-handed, also have right-handed components. This surprising experimental fact points toward another way of constructing a gauge theory of the weak interaction—this time an even more fascinating one, namely, a unification of the weak and electromagnetic interactions. We are aware that electromagnetic interaction takes place with both left-handed and right-handed fermions. Since neutral-current interactions occur in a similar fashion, it is not a farfetched idea to combine these interactions in one theory. (Ideas along this line were developed between 1958 and 1967 by Sheldon Glashow, Abdus Salam and John Ward, and Steven Weinberg.) Everything

Figure 18.1
Theorist Sheldon Glashow (left) and Steven Weinberg together
on a special occasion—a press conference at Harvard University
on the day when it was announced that they and Abdus Salam
had received the 1979 Nobel Prize in physics. Sheldon Glashow
has played a significant role in the development of particle physics since 1960 by stimulating new directions in research. He holds
a chair in theoretical physics at Harvard.

Steven Weinberg's important contributions to theoretical physics are mostly in field theory. He was among the first to apply
field theoretic methods to weak and strong interaction physics.
Furthermore he was and is actively engaged in research on gravity theory. He is currently professor of theoretical physics at the
University of Texas in Austin.

The Unified Theory of Electroweak Processes

Figure 18.2
Theorist Abdus Salam, who shared the 1979 Nobel Prize in Physics with Sheldon Glashow and Steven Weinberg for his contribution to the development of weak interaction theory, a field in which he had been interested since the late 1950s. He published some of his important contributions jointly with John Ward. Abdus Salam holds a chair in theoretical physics at Imperial College in London and is director of the International Center of Theoretical Physics in Trieste (Italy).

we know about the weak interactions agrees with the simplest version of such a theory, and chances are that it is at least a good approximation of the "correct" theory (figures 18.1 and 18.2).

The Unification of Electromagnetic and Weak Forces

A unified theory of electromagnetic and weak interactions must be able to account for four particles: the two W particles, the Z particle, and the photon. This is why we require a group with four different charges. The simplest way of constructing such a group is to take the previously discussed weak isospin group SU(2) and add to it another group, one that has only one charge. Mathematicians call such a group U(1). The total gauge group is then a product: SU(2) × U(1).

We can now calculate the masses of the gauge particles by using the mechanism of spontaneous symmetry breaking. Doing so, we come upon a remarkable phenomenon: the two W particles acquire some mass, the third neutral particle acquires a slightly larger mass, and the fourth particle remains massless.* Nothing seems more natural, therefore, than to identify the last particle as the photon. Having done so, we arrive at a unified theory of the weak and electromagnetic interactions. In this theory, the strength of the coupling of W particles with leptons and quarks is comparable to the strength of electromagnetic coupling.

Moreover, we are now in a position to predict the form that the neutral-current interaction takes. Since we have a unified scheme that includes the electromagnetic interaction, neutral-current interactions are not purely left-handed any more (as they were when

---

* For the time being, we must assign an arbitrary value to these masses because the absolute scale of the mass can be determined only experimentally.

our theory dealt only with the weak interaction). Neutral-current interactions depend on only one parameter, which is a free parameter of electroweak theory. This parameter, whose value has to be determined experimentally, is called the weak interaction angle Θ_w. This angle describes how strongly the weak and electromagnetic interactions are linked to each other. When Θ_w equals zero, there is no relationship whatsoever. Recent experiments have determined that Θ_w is between 27 and 29°.

This theory of flavor dynamics allows us to predict that the mass of the W bosons must be approximately 80 GeV and that of the Z particle around 90 GeV (these masses depend on Θ_w). To actually discover these particles in an experiment would be the ultimate test of our theory, and the discovery of these particles constitutes one of the major tasks for high-energy physics today.

All Those Other Particles

So far we have discussed only the lightest leptons and quarks. What about the rest? How do they fit into our scheme? Well, we just add more left-handed doublets of the weak isospin as follows:

$$\begin{pmatrix} \nu_e \\ e^- \end{pmatrix} \quad \begin{pmatrix} \nu_\mu \\ \mu^- \end{pmatrix} \quad \begin{pmatrix} \nu_\tau \\ \tau^- \end{pmatrix}$$

This simple assumption is in excellent agreement with all experimental data. The muon and τ systems behave exactly as the electron system does.

We proceed in exactly the same manner for quarks.

To find places for the strange and charmed quarks, we simply add another doublet of the weak isospin:

$$\begin{pmatrix} u \\ d' \end{pmatrix} \quad \begin{pmatrix} c \\ s' \end{pmatrix}$$

This procedure introduces one slight complication, however. Notice that we have put a prime symbol on the d and the s. This is done for the following reason. The u quark is coupled chiefly to the d quark by means of the weak interaction, but not entirely so. There is a 5 percent probability that it is coupled to the s quark. By analogy, the charmed quark is coupled chiefly to the s quark, but also has a 5 percent chance of coupling with the d quark. This phenomenon, which is called weak interaction mixing, is highly important. If it did not occur, the strange particles would be coupled by the weak interaction only to the heavy charmed quark; that is, the strange particles could not decay. They would be stable. (We know, however, that s quarks do decay.) In the above scheme, therefore, the weak interaction mixing is denoted by writing d' and s' to remind ourselves that the u quark is coupled to the d quark only 95 percent of the time (and the same, of course, for the c and s quarks).

One question the reader may ask is how the b quark fits into this scheme. All we can say in reply is that right now we do not know. The weak-interaction properties of the b quark have not yet been studied in detail. However, the very existence of the b quark means that there must be several new mesons consisting of the b quark and one of the "light" antiquarks (for example, the mesons with electric charge -1 and quark composition $\bar{u}b$). These mesons are expected to have a mass of 5.2 GeV. If we wish to determine the weak-interaction properties of the b quark, we must first find these mesons (called B mesons) and study their weak decays.

One More Quark

In the absence of unequivocal information about the nature of the b quarks, we have no choice but to speculate. If we take another look at the lepton and quark schemes we just outlined, we note an odd asymmetry between them. We have three lepton doublets but only two quark doublets. We find ourselves in the same bind as when we came upon the muon and its neutrino and had two lepton doublets but only one quark doublet. We restored the symmetry at that point through the introduction of the new charmed quarks that form a weak-interaction doublet with the strange quarks. We can resolve our current predicament, too, by adding yet another quark flavor. This one we call the t quark, and, together with the b quark, it forms a weak doublet (t and b signify top and bottom or, if you prefer, truth and beauty):

$$\begin{pmatrix} u \\ d' \end{pmatrix} \quad \begin{pmatrix} c \\ s' \end{pmatrix} \quad \begin{pmatrix} t \\ b' \end{pmatrix}$$

Now we have restored the symmetry between leptons and quarks. There are three pairs of each. However, we note that this analogy between them is merely that and nothing more. We are merely following a hunch, the kind of hunch that has helped us previously. There is no profound underlying theory that insists that quarks and leptons must always come in left-handed doublets.

What is there to say about the new t quark? First of all, we have no evidence that it even exists. If it does, though, it has to be much heavier than the b quark, which has an effective mass of about 5 GeV. In fact, the experiments indicate that the t quark must be heavier than 18 GeV. It will be an important task for future experiments to prove or deny the existence of the t quark.

The t and b quarks are related to each other through the weak interaction. However, the b quark is lighter than the t quark, and therefore the b quark cannot decay via the weak interaction into a t quark. The situation resembles the c-s quark case, in which the strange quarks decay via weak interaction mixing. By analogy, we assume that some weak interaction mixing occurs between the b quark and the two other quarks with charges of $-1/3$ (d and s) and that the b quark also decays through such mixing. There is no prediction from first principles of how strong this mixing is. However, various constraints tell us that it cannot be large. In addition, these constraints indicate that the dominant weak decay of the b quark proceeds via the c quark. The b quark decays by emitting a c quark and a virtual W^- boson, the latter producing whatever it can at the corresponding energy, either a quark pair, such as $\bar{u}d$ or $\bar{c}s$, or lepton pairs, such as $\bar{\nu}_e e^-$, $\bar{\nu}_\mu \mu^-$, or $\bar{\nu}_\tau \tau^-$.

In the summer of 1980, physicists working at the electron-positron storage ring at Cornell University discovered effects that have been interpreted as the consequence of the weak decay of the B mesons. These discoveries agree with the theoretical prediction that the decay of b quarks results mostly in the production of c quarks. The Cornell physicists were also able to obtain information about the mass of the B mesons. The mass of these particles, it turns out, is about 5.26 GeV, which is in very good agreement with theoretical predictions using the γ mass as an input.

The above ideas about the weak-interaction properties of the b quark depend on the existence of the t quark. Unfortunately, we have no way of calculating at what level of energy the t quark may make its appearance. It may well be that many years will pass before it is discovered, and so this, too, remains one of the important challenges of contemporary physics. Various theories predict the effective t mass to be 18—20 GeV. If this is true, the t-quark effects will be discovered before 1984 at DESY.

Discovery of the W Boson

The ideas discussed here were dramatically confirmed at CERN in January 1983. Investigating proton-antiproton collisions at high energies, the CERN physicists reported evidence for the W boson, which is produced by quark-antiquark annihilation and decays shortly afterward into an electron or positron and a neutrino. Thus far ten events of this kind have been observed. On the basis of these events the W mass is estimated at about 83 GeV, in very good agreement with the theoretical prediction.

XIX

Does Physics Come to an End?

Contemporary particle physics achieved an extraordinary breakthrough with the development of first a theory for the strong interaction and then a theory for the combined electroweak interactions. All phenomena that fall in these domains can be explained by those two theories.

The electroweak theory, however, contains a serious and as yet unresolved problem. It fails to explain why the electric charges of leptons and quarks are quantized in definite units. (The electric charge of electrons and muons is −1, and the electric charge of quarks is either +2/3 or −1/3). It appears as though physics contains a secret law that compels the various particles to contain only well-defined charges. But what is that law?

The SU(2) × U(1) framework for the electroweak

interaction has two neutral charges: the U(1) charge we discussed in the previous chapter and the neutral weak isospin charge. Electric charge, while it is a certain combination of these two, still depends on a free parameter, namely, on the weak mixing angle Θ_w. This implies that electric charges are not quantized, but arbitrary. For example, we can arrange it so that the electric charge of the quarks is some arbitrary number—$(2/\pi)$ instead of $(2/3)$, for example. To arrive at an understanding of the quantized charges, we have to go beyond the electroweak theory. We shall explain shortly how we might go about doing so, but first we have to mention one further problem.

Unifying the Strong and Electroweak Theories

We have succeeded within the SU(2) × U(1) framework in combining the electromagnetic and weak interactions. However, this leaves the strong force unaccounted for, and we must ask whether it is possible to construct a theory of both strong and electroweak interactions. Such a theory would have to explain not only the structure of strong interactions (that is, the color forces), but also their strength. We mentioned previously that the coupling strength of strong interactions is quite large relative to that of electroweak interactions. The QCD fine structure constant relevant for strong interactions at the energy level of a few GeV is on the order of 0.2, whereas the fine structure constant of electromagnetism is merely $1/137$. Strong interactions therefore are indeed far stronger than electromagnetic ones. This fact, too, must be explicable within a unified theory of the interactions.

How can we construct such a theory? Let us cast another glance at the gauge groups of QCD and of the weak and electromagnetic interactions: SU(3) × SU(2) × U(1). One way of constructing a unified theory is to try to embed these three groups in a larger overall group. Such an endeavor presents no problem for mathematicians, at least in principle. However, constructing the fermions (electrons, neutrinos, quarks) is another problem. We have relatively little choice in the matter of fermions: they are either color singlets (leptons) or color triplets (quarks). It turns out that only relatively few groups can contain group SU(3) as well as group SU(2) × U(1) and simultaneously provide a correct account for the spin 1/2 particles. The first group that includes the smaller groups SU(3) and SU(2) × U(1) and also provides an accurate description of the fermions is SU(5), as noted by Howard Georgi and Sheldon Glashow in 1974. To see how this works, we first take a look at the lightest fermions, that is, the electron, its neutrino, the u and d quarks, and their antiquarks. All these fermions, including the antifermions and the three colors of the quarks, can be grouped in the following manner:

$$\begin{pmatrix} \nu_e \\ e^- \end{pmatrix} \quad e^+ \quad \begin{pmatrix} u\,u\,u \\ d\,d\,d \end{pmatrix} \quad (\bar{u}\,\bar{u}\,\bar{u}) \quad (\bar{d}\,\bar{d}\,\bar{d})$$

We notice that there are fifteen fermions altogether, which we now divide into the following two systems:

$$\begin{pmatrix} \nu_e \\ e^- \end{pmatrix} \bigg| \ \bar{d}\,\bar{d}\,\bar{d} \bigg) \quad \begin{pmatrix} u\,u\,u \\ d\,d\,d \end{pmatrix} \bigg| \ \bar{u}\,\bar{u}\,\bar{u} \ \bigg| \ e^+ \bigg)$$

The first system contains five fermions, the second

ten. It turns out that precisely these fermion systems are needed in order for the large symmetry group SU(5) to make sense (they are "representations" of SU(5)). It is also worth noting that we obtain precisely the leptons and quarks we need, neither more nor less. The electric charges of the leptons and quarks are no longer arbitrary. To see that this is so, all we need do is glance at the first of the two systems. Group SU(5) includes twenty-four different charges altogether. (The larger the group, the more charges it has.) Group SU(2) has three charges, group SU(3) contains eight charges, and in general group SU(n) has $n^2 - 1$ charges. One of these twenty-four charges must be the electric one.

The charges of all elements in a representation *must* add up to zero, and so the electric charges of the five fermions grouped together above must add up to zero. Since the electric charge of the neutrino is zero, we now come upon a relationship between the electric charge of an electron and that of the $\bar{d}$ quark: $Q(e^-) = Q(d/3)$. In this way we obtain precisely the electric charge observed in nature. By doing the same thing with the ten-member group of fermions, we obtain the desired electric charge of 2/3 for the u quark. Within SU(5), therefore, the electric charges are quantized and the theoretical results coincide with what we find in nature.

In addition, we are in a position to do some theoretical extrapolations and calculate the QCD fine structure constant as well as the electroweak mixing angle. The results of these calculations are

$$\theta_w \approx 38°$$
$$\alpha_s \approx \frac{8}{3}\alpha \approx \frac{1}{50}$$

These two values, however, totally disagree with what we found in our experiments. The QCD fine structure constant that is relevant for the energy region of a few GeV is about seventeen times greater than the value extrapolated theoretically. The weak mixing angle is approximately 27°, not 38°, in our experiments. This presents a major obstacle to the immediate acceptance of the SU(5) group theory.

Now that we have mentioned one problem, we must confess to yet another. This problem has to do with the expected life of the proton. The five- and ten-member groups of fermions contain both leptons and quarks. The gauge theory based on group SU(5) contains twenty-four gauge particles, including the eight gluons of QCD, the two W bosons, the Z boson, and the photon. The remaining twelve bosons are entirely new and mediate as yet unknown interactions. These interactions are rather peculiar, however. They can transform a lepton into a quark and vice versa, not a surprise in a theory unifying leptons and quarks. As a result, though, a proton can decay into leptons and mesons. Specifically, a proton can turn into $e^+ \pi^0$ (a positron and a neutral meson). This of course presents a serious problem for SU(5), for experiments tell us that the proton's lifetime is at least 10^{30} years (as discussed in chapter 4).

Since the calculated lifetime of the proton in the SU(5) gauge theory depends on the mass of twelve new gauge particles, we can assign a lower limit to the mass of these bosons, which still turns out to be enormous: 10^{15} GeV. Consequently, the unification of the strong, weak, and electromagnetic interactions occurs only when enormous masses are involved.

Does Physics Come to an End?

A Tremendous Mass Scale

The appearance of such a scale of energy and mass has an interesting effect. Let us take another look at the coupling strengths of the various interactions, which we must now consider at 10^{15} GeV or higher because they are manifestations of the same underlying unified theory. For example, the SU(2) × U(1) mixing angle is expected to be 38° at these energies. However, our phenomenological information about the strengths of these interactions tells us that Θ_w is only about 27°. Perhaps the reason for this discrepancy is simply that all our experiments so far have been run at relatively low energies, low, at least, relative to 10^{15} GeV. How, then, are we to establish a connection between the prediction of the grand unified theory for the various coupling strengths and the observed coupling strengths at lower levels of energy?

The picture that physicists have in mind is something like this. Once we consider physical phenomena above and beyond 10^{15} GeV, the mass scale for the grand unification of the interactions, we can no longer tell the difference between strong, weak, and electromagnetic interactions. Only one type of unified interaction is observed. Even the difference between leptons and quarks disappears—they are just manifestations of one and the same type of underlying basic fermion. However, as soon as we dip below 10^{15} GeV, the different interactions develop lives of their own and we can distinguish between them and between leptons and quarks. We can watch the development of the different coupling strengths according to the laws of QCD and the laws of electroweak interactions.

For example, we can calculate the coupling strength of QCD as a function of the grand unified mass scale; it is dependent on the energy at which the unification of all interactions sets in. And we can also do the opposite. Since we know the various coupling strengths from our experiments, we can calculate the mass scale at which the grand unification should set in. The outcome is surprising—or perhaps not, since theory has served us in such good stead so many times before. We find that the energy is 10^{15} GeV, the same energy we obtained as the lower limit for the unification theory SU(5). It is unclear at present whether this is merely a coincidence or if something like the SU(5) scheme actually applies.

Let us suppose that it does, though, in which case we expect the proton to be unstable and its lifetime to be just a little bit longer than what has been established experimentally, namely, between 10^{30} and 10^{32} years. Sophisticated new experiments are being prepared to provide us with a new limit on the lifetime of the proton or to establish the existence of its decay. The basic idea of these experiments is to monitor the nucleons of a very large mass of material (the larger the mass, the greater the number of protons and neutrons and hence the greater the likelihood of observing the decay). These experiments are being conducted deep underground in order to suppress background from cosmic rays.

Since the experiments require large amounts of material, the material must be cheap and plentiful like water, concrete, or iron. The particles emitted in the hypothetical proton decay are expected to have relatively high energies. Their speed, though less than that of light in a vacuum, will nonetheless be higher than that of light in water. (The speed of light in a vacuum is what we usually mean when we speak of the speed of light. It will be just a bit less in a medium like water.) An electrically charged particle traveling through water at the speed of light emits a special kind of radiation called Cerenkov radiation, after the Soviet physicist who explored it in great detail in the 1930s. (The Cerenkov effect is the optical analog of the sonic boom generated by an airplane traveling faster than the speed of sound.) By measuring Cerenkov radiation, we can obtain information about the energies emitted in proton decay. For example, proton decay may produce a positron and a neutral pion. The latter subsequently decays into two photons, each of which produces an electromagnetic shower, and subsequently Cerenkov radiation.

An experiment monitoring large amounts of water for signs of proton decay is being conducted in the Morton salt mine east of Cleveland by a group of physicists from the University of California at Irvine, the University of Michigan at Ann Arbor, and Brookhaven National Laboratory. There 10,000 tons of water are being investigated at a depth of 600 m (figure 19.1). Other experiments are being performed or will be performed at various other places in the United States, the Soviet Union, Japan, and Western Europe. In particular we should like to mention an experiment which started in the summer of 1982 in a cave close to the Mont Blanc tunnel connecting France and Italy, and an experiment which will start in 1983 in the Frejus tunnel between Grenoble and Torino.

For several years a group of Indian and Japanese physicists have conducted a search for proton decay

Figure 19.1
The cave of the Cleveland proton-decay experiment before it was filled with water. The cave contains 10,000 tons of water which are monitored continuously by a large array of photomultipliers. The latter are used to detect the Cerenkov radiation caused by the particles emitted in a proton decay (courtesy Irvine-Michigan-Brookhaven collaboration).

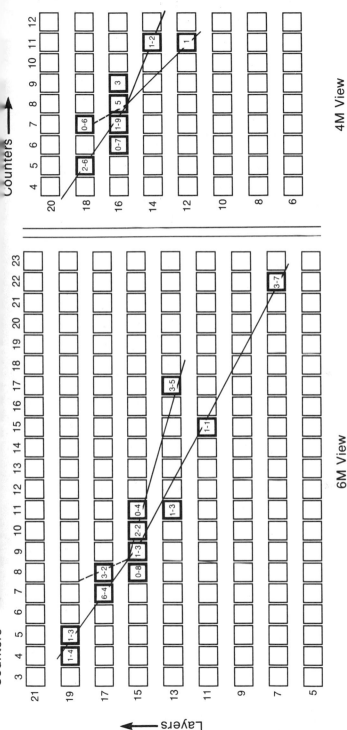

Figure 19.2

Event 587, observed by an Indian-Japanese scientific team in the Kolar gold mine on November 10, 1981, is consistent with what we expect to be happening during proton decay, that is, the debris of a proton, decaying into a positron (track leaving upward), and a neutral pion decaying immediately after its creation into two photons (the two tracks going downwards).

in the Kolar gold mine in southern India at a depth of 2300 m. Since November 1981 three events have been observed which can be interpreted as due to the decay of the proton. One of these events is shown in figure 19.2. However, the findings of the Indian-Japanese collaboration are being challenged by other physicists, so the scientific world has to wait for further experimental results in order to reach a firm conclusion. Nevertheless, if the three events found in the Kolar gold mine are interpreted as being due to proton decay, a statement can be made about the stability of the proton: it lives 6×10^{30} years.

It is impossible to overestimate the importance of proton decay for the future development of physics. Finding out that all matter eventually decays, leaving a debris of photons, positrons, neutrinos, and so on, would certainly be a highly significant discovery (and something of a shock for many). However, not to worry. Even if the proton turns out to have a lifetime of "only" 10^{31} years, it is for all intents and purposes stable. The earth would loose the minute amount of 0.001 gram of matter per annum, and the chance of just one nucleon decaying in someone's body during an average human lifetime is only about 1 percent.

Back to the Baryon Number

If the proton ultimately decays into leptons and photons, this means that the baryon number is not conserved, which may provide the solution to the puzzle of why nature is rich in matter but devoid of antimatter. If the baryon number is conserved, the number of baryons in the world is constant, which implies that the universe began with a nonzero baryon number (and a rather high number at that, with the number of baryons existing now being the same as then).

Matters, however, are quite different if the baryon number is not conserved, especially if the violation of the baryon number conservation law proceeds via the new types of boson interaction we discussed for the SU(5) scheme. These interactions are essentially negligible at low energies, that is, at the energies at which physicists carry out their experiments. In the early stages of the universe, however, the relevant energies may have been very large indeed (on the order of the grand unified mass scale or even larger), in which case the new boson interactions, which are the cause of the instability of the proton, do become relevant and may be responsible for the generation of baryons in the universe. If something decays, it can also be produced. Thus the idea of a unification of the interactions provides us with a possible explanation for the nonzero baryon number in the universe.

Keeping these new interactions in mind, we might even be able to calculate the number of baryons in the entire universe. Several attempts to follow this line of thinking have been made, and although the calculations are still fraught with uncertainties, physicists have come up with a number whose magnitude is consistent with the observed baryon density in the universe. The exact number, however, depends on parameters that are as yet unknown. Still, what has been found so far is extremely interesting, and the future will show whether physicists are on the right track or not.

I should like to note another success of the SU(5)

scheme. I mentioned above that the predicted value of the electroweak mixing angle is 38°. This angle is relevant for physics only at or above 10^{15} GeV, however. To be able to say something intelligent about the electroweak angle at the energies at which our experiments are conducted (about 100 GeV), we must extrapolate the electroweak coupling strength down to 100 GeV. This can be done within the framework of $SU(2) \times U(1)$ theory, and it turns out that the extrapolated weak angle at such comparatively low energies is $\Theta_w = 27°$, which agrees nicely with the experimental observations. This may be regarded as another indication that there is at least something like a unified theory of the various interactions and that such a theory becomes relevant only at the very high energies of 10^{15} GeV.

Other Unification Theories

We have discussed the SU(5) theory as only one example of a unified field theory. We do not want to imply that we give it special preference over any other theory. In recent years, many theoreticians have worked on other schemes, one interesting example being a theory based on gauge group SO(10), the orthogonal group in ten dimensions. In this theory, which has been proposed by Howard Georgi, Peter Minkowski, and me, the u and d quarks and the electron and its neutrino are represented by a sixteen-member lepton-quark system called the 16-dimensional representation of SO(10).

This sixteen-fold system of particles consists of:

$$\begin{pmatrix} \nu_e & \vdots & u\,u\,u & \vdots & \overline{u}\,\overline{u}\,\overline{u} & \vdots & \overline{\nu}_e' \\ e^- & \vdots & d\,d\,d & \vdots & \overline{d}\,\overline{d}\,\overline{d} & \vdots & e^+ \end{pmatrix}$$

Note that the three colors of the quarks are taken into account explicitly. In the SO(10)– scheme one is dealing with sixteen fermions, that is, one more than in the SU(5)– scheme, which contained only fifteen fermions. The additional fermion is denoted here by $\overline{\nu}_e'$, a new type of neutrino, which is supposed to exist, but whose existence is not yet confirmed.

The symmetry, present in the SO(10)– scheme, is larger than the symmetry of the SU(5)– scheme: all sixteen fermions are related to each other by the large symmetry. Otherwise its predictions for particle physics—for example, for proton decay—are very similar to those of the SU(5)– theory. There is one difference, however, that may be of great importance. In the SO(10) theory one expects the neutrinos to have small mass of the order of 1eV, with all the aforementioned consequences for particle physics and cosmology.

So far we have discussed the unification of only electroweak and strong interactions. We have not discussed gravity, and the question arises as to how gravity fits into our scheme. Even Einstein spent many years of his life trying to combine the theories of electrodynamics and gravity, but without success. The reasons for his failure are easy to understand today. During his lifetime pitifully little was known about the strong and weak interactions. Any attempt to combine electromagnetism with gravity while leaving out the other interactions is bound to fail. Today we know the structure of the strong and electroweak interactions, and it turns out that all of them can be described by gauge theories similar to Maxwell's theory of electrodynamics. For this reason one is tempted to revive Einstein's program and to construct a unified theory of all interactions, including gravity.

Although many attempts to solve this problem are being made at present, none has succeeded so far. We are dealing with an especially knotty problem in this case, the more so because gravitational interaction differs from other interactions in one significant respect: gravity is an interaction that has an intrinsic scale. This scale is determined by the gravitational constant in Newton's law of universal gravitation, which describes the strength of the interaction of the gravitational attraction between the earth and, for example, an apple about to drop from a tree.

For elementary-particle physics it is useful to translate the gravitational constant into an energy scale by means of quantum theory. Doing so, we come upon a very large energy $1.1 \cdot 10^{19}$ GeV, which is the Planck energy (it corresponds to a characteristic length of 10^{-33} cm). The Planck length is a parameter, which is determined by Newton's constant and by Planck's constant $\hbar$. Physicists believe that at distances of the order of the Planck length our conventional picture about the structure of space and time breaks down. Nobody knows what really happens. We cannot help but note that the Planck energy is not that much greater than the energy of 10^{15} GeV we obtained for the unified theory. It may well be that the energy scale that unites the strong and weak interactions is directly related to the Planck energy, and therefore to gravity.

The construction of a unified theory of gravity and of the strong and electroweak interactions presents a

Does Physics Come to an End?

major challenge for theoretical physicists in our time. The future will show how gravitational interactions mesh with the scheme of particle interactions.

Our discussion of the grand unification of the interactions has not touched so far on one particular problem: the rather plentiful number of leptons and quarks. As we have seen, we can construct grand unifying theories that unite the strong and electroweak interactions. Such theories explain the various observed interactions—their strengths, the value of the mixing angles, and the patterns of quarks and leptons (quarks as color triplets, leptons as color singlets). They fail to tell us, however, how many quarks or leptons there are in this world. So far we have mentioned the fifteen basic light fermions that are incorporated into the SU(5) grand unification scheme. Besides these fermions, there are many others, heavier ones. As a matter of fact, the muon and its neutrino and the charmed and strange quarks compose another fifteen-member unit, and the same can probably be said of the t lepton, and of the t-b quark system.

Fermions appear to crop up as families. The lightest set of fermions (electrons, electron-neutrinos, and u and d quarks) build the first generation, which is succeeded by the muon, muon-neutrino, charmed and strange quarks, and so forth. How many generations are there, and why do they exist? At present we do not know. All we can say is that the world would not look all that different if the only thing that existed was the first generation of fermions. Everything that we can observe in our macroscopic world—its galaxies, stars, the earth, a tree, ourselves—consists of the first generation of fermions, namely, the u and d quarks (nucleons) and the electrons. The fermions of the second and higher generations build up new kinds of heavy matter, which can be produced in high-energy physics laboratories and which decays quickly. We can describe the three known generations as follows:

First generation	Second generation	Third generation
$\begin{pmatrix} \nu_e & \vdots & u \\ e^- & \vdots & d \end{pmatrix}$	$\begin{pmatrix} \nu_\mu & \vdots & c \\ \mu^- & \vdots & s \end{pmatrix}$	$\begin{pmatrix} \nu_\tau & \vdots & t \\ \tau^- & \vdots & b \end{pmatrix}$

Why the new fermion generations exist at all is still a mystery. Moreover, we do not know how many more generations of fermions, leptons, and quarks there are besides those noted above. It may be that Nature is content with the three generations we have discovered so far, but it may also be that there are many more generations, perhaps infinitely many. A complete theory of elementary particles and their interaction will give us the answer.

At present we can imagine two possible ways for physics to develop in the future:

1. The various quarks and leptons are part of a large system that is based on a gauge theory. If this is the case, then quarks and leptons are truly elementary particles.
2. The world of atoms and nucleons manifests an astonishing richness of structure. There exist many, in fact infinitely many, different atomic and nuclear states as a result of the existence of the several constituent particles and their interactions. That way one is able to explain an infinite number of states on the basis of a comparatively limited number of constituent particles. The same principle that, based on the theory of quarks, helps explain

the existence of the many kinds of hadrons, may also help us with leptons and quarks. Perhaps there is one more level of structure, and leptons and quarks are composite systems consisting of as yet unknown subunits. Such lepton-quark subunits have already been introduced into the physics literature, bearing such names as subquarks, preons, stratons, rishons, and haplons.

So far, physicists have discovered four different kinds of subunits for matter, as shown in the following scheme:

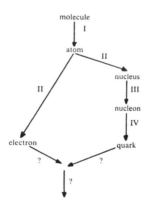

There may be even more levels in the hierarchy of substructures, nothing is known at present. Physicists have examined leptons (electrons, muons) for substructure, but so far have found no evidence indicating any. The electron and muon seem to be pointlike (structureless) particles up to a distance of 10^{-17} cm. In other words, if they have a substructure it appears only at distances smaller than 10^{-17} cm. Moreover, quarks have the peculiar characteristic that they cannot be produced freely but are permanently confined within color singlet hadrons. It seems an odd notion, then, to suppose that they are composed of yet smaller units. The permanent confinement of quarks may be a sign that physicists have reached the end of the road in exploring the structure of matter.

If the leptons and quarks are composite objects, we expect there to be far more leptons and quarks than we know about now, for example, exotic leptons and quarks with spin 3/2 and 5/2 and more. It remains to be seen whether new experiments will find such exotic objects.

The question whether leptons and quarks have a substructure (that is, are composite objects) is intimately related to the question of whether there is a final theory for matter. If quarks and leptons turn out to have subcomponents, we can easily envisage this game being repeated over and over, ad infinitum. In principle, there could exist an infinite number of levels of fine structure, and the field of exploration would be endless.

If quarks and leptons are the ultimate constituents, however, physicists expect to be able to construct a final theory. Right now we do not know how quickly this might be accomplished, but we do expect that someday there will be formulated a theory for all of particle physics that will follow the line of thinking we have sketched out above. Such a theory, of course, will have to account for the number of leptons and quarks, their masses, their patterns of interaction, the masses of the W and Z bosons, and other properties. This theory will have only one parameter, the Planck energy which is provided by the gravitational interaction and which, as it were, provides the scale for energy and mass in the universe.

Any physicist claiming to have such an overall theory would have been regarded as quite mad until recently. After the rapid developments in elementary-particle research in the 1970s, however, the mad hatter may turn out to be a genius. For some physicists, the end of the road is in sight, and the main question is simply, How long until we reach it?

XX

The Future of
the High-Energy
Physics Program

In the last chapter we risked the suggestion that the end of physics might be in sight. Of course, new data or new theoretical insights may invalidate this suggestion. It would not be the first time that physics has regarded its estate with such naive optimism only to be caught short by a new development. For example, the young Max Planck was told by his physics professor in Munich that he should drop the subject, it was nearly wrapped up—that was just before the quantum and relativity theories shook physics to its roots.

Therefore, it could well be that our current view of subatomic physics will need to be revised. For this reason, it is exceedingly important to test our ideas experimentally in particle collisions as higher energies become available to us. Various new colliders are under construction in nearly every part of the world—to accelerate protons and antiprotons, electrons and positrons to ever higher energies.

In the United States, preparations are under way to construct a proton-antiproton collider at FNAL which would allow the study of frontal collisions of protons and antiprotons with energies of 1000 GeV each. Physicists at SLAC, following an ingenious proposal, intend to build an electron-positron collider to produce the Z particle—provided it exists. In the Soviet Union, physicists are planning to construct a proton accelerator that will accelerate protons to energies of up to several thousand GeV.

The physics community in the United States is considering the construction of a large proton-antiproton collider in the desert of New Mexico. This ring would have a radius of about 30 km and would allow the acceleration of protons and antiprotons to 20 000 GeV. This accelerator, the cost of which is estimated to be about one billion dollars, would be the first one to reach energies much above the energy scale given by the weak interactions. It may well be that entirely new phenomena will show up once these energies are reached.

In view of the current state of physics, several large projects in Western Europe are of particular interest, for they will be well suited to the study of the details of weak interactions and of chromodynamics at high energies. At CERN, construction has begun near Geneva on a large electron-positron collider called LEP (figure 20.1), and DESY in Hamburg wants to con-

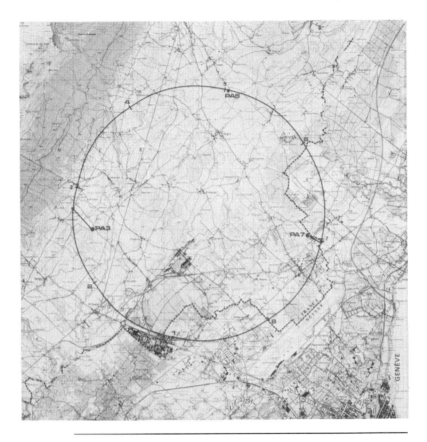

Figure 20.1
The LEP storage ring west of Geneva, Switzerland. The ring will
be constructed below ground and will be invisible from the surface. The plan is to construct eight experimental halls here denoted by P1 . . . P8. The western part of the ring runs through the
Jura Mountains. Geneva airport is visible in the right-hand cor-
ner. LEP runs right through CERN, here denoted by 9 (courtesy
CERN).

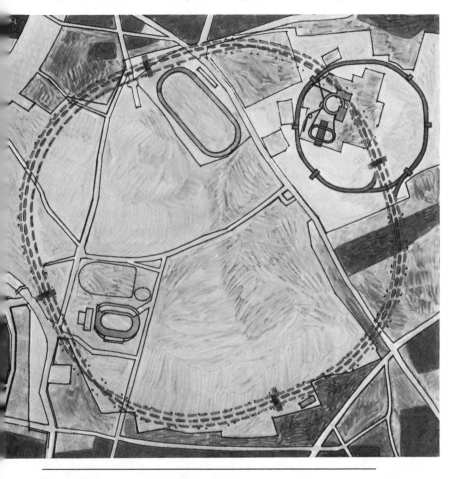

Figure 20.2
A schematic drawing of the planned HERA machine at DESY, Hamburg. On the left is the DESY site with the PETRA ring (courtesy DESY).

struct HERA, a powerful electron-proton collider (figure 20.2).

The construction of LEP is the most ambitious of these projects. Due to its high costs (about 500 million dollars) LEP can be built only with the joint participation of all Western European countries. Using it, physicists hope to achieve energies of 140 GeV eventually (that is, a total of 280 GeV from both beams) in electron-positron collisions. LEP, therefore, is an ideal machine for studying the structure of the weak interactions in which new phenomena, such as the production of the W and Z particles, should manifest themselves.

LEP will be the largest accelerator built thus far in the world (figure 20.3). The storage ring will have a circumference of nearly 30 km and will be constructed in several phases. Once the first phase is completed, experiments will be able to achieve energies of up to

Figure 20.3

An artist's view of an interaction region at the LEP ring. The big hall accommodating the particle detectors must be built about 80m below the surface since the LEP ring itself will be constructed in a tunnel about 80m below the surface (courtesy CERN).

60 GeV, sufficient to produce the Z boson (current theory predicts a 90 GeV energy for it).

HERA is, strictly speaking, a double storage ring, one for protons, the other for electrons. The proton ring can accelerate protons up to 800 GeV, and the electron ring will reach an energy of 30 GeV.

The reader may well ask why the energy of the protons in HERA is planned to be is twenty times higher than that of the electrons. There is a good reason for this. We know that an electron colliding with a proton at high energies will react with one of the quarks inside the proton. Strictly speaking, HERA is not an electron-proton collider but a quark-electron collider. For experimental purposes, electrons and quarks should have similar energy. However, we know from our previous discussion that the quarks inside a fast-moving proton carry only a fraction of the total energy. Therefore it is useful to have the 30 GeV electron colliding with protons of a much higher energy.

HERA will be an ideal machine for studying the interactions of electrons with quarks at very high energies. Specifically, we will be able to study the predictions of QCD and obtain confirmation of the structure of the weak forces. The energies in electron-quark collisions at HERA will be so large that the weak and electromagnetic forces will become comparable in strength.

Using LEP and HERA, physicists will therefore be able to study the dynamics of quarks and electrons at very minute distances, and we shall be able to see whether these particles are structureless objects (down to distances of about 10^{-17} cm, ten thousand times smaller than the diameter of the proton). If electrons and quarks have subunits, this might be one way of discovering them.

As in the fields of astrophysics and astronomy, the research in high energy physics, using powerful accelerators, has led to a large number of exciting discoveries during the last twenty years. The view of physicists regarding the structure of matter and the architecture of the universe at very small distances has changed entirely during those years. Like the exploration of space, high energy physics research is basic research, that is, research carried out without direct reference to a specific technological goal. Nevertheless many offspring technological developments have started in the high energy physics laboratories—for example, the use of very powerful computers for handling vast amounts of data, the development of sophisticated electronic devices, and the large-scale use of superconducting magnets. The building of powerful high energy accelerators is a considerable challenge to engineers and scientists alike. It is this collaboration that stimulated the technological breakthroughs in the past and, no doubt, will stimulate new ones in the future.

In spite of all that, the main goal of high energy physics research is the study of matter under very extreme circumstances. What can be more fundamental than to investigate the structure of matter at very small distances, to find out what matter really is, why there is matter at all, and what happens to it in the future.

Epilogue

I hope the reader of this book has obtained some insight into the current ideas of physicists regarding the structure of matter in our world. Admittedly, a deep understanding of those ideas requires a detailed knowledge of quantum mechanics and theory of relativity, including the mathematical technology that goes with it. I believe, however, that the basic ideas of modern physics are quite simple and are within the reach of laymen. I hope this book contributes to the goal of making modern science more understandable for the non-experts.

Furthermore, I hope that the reader has gotten a sense of the relevance of basic research. Unlike research in applied science and industry, research in particle physics is motivated solely by our curiosity about nature, our desire to further explore the subtlety and symmetry of physical laws. The insights developed through this research will be remembered even in the distant future.

Appendix

Leptons and Quarks

The values in brackets are the rest masses of leptons and u quarks given in GeV. There is no mass for the neutrinos because it is as yet unclear whether neutrinos have mass or not. Quarks are not observed as free particles. Therefore, we use their effective mass, which we introduce in order to understand the masses of the particles composed of quarks.

I. $\begin{pmatrix} \nu_e \\ e^- [0.00051] \end{pmatrix}$ $\begin{pmatrix} u\ [0.3] \\ d\ [0.3] \end{pmatrix}$

II. $\begin{pmatrix} \nu_\mu \\ \mu^- [0.106] \end{pmatrix}$ $\begin{pmatrix} c\ [1.5] \\ s\ [0.45] \end{pmatrix}$

III. $\begin{pmatrix} \nu_\tau \\ \tau^- [1.78] \end{pmatrix}$ $\begin{pmatrix} t\ [?]^* \\ b\ [4.9] \end{pmatrix}$

*The t quark has not been discovered yet. Its mass must be larger than 18 GeV.

Appendix

Important Mesons and Their Quark Structure

Meson	Structure	Mass
π^+	$\bar{d}u$	0.140 GeV
π°	$\bar{u}u/\bar{d}d$	0.135
π^-	$\bar{u}d$	0.140
K^-	$\bar{u}s$	0.494
$\bar{K}^\circ$	$\bar{d}s$	0.498
D°	$\bar{u}c$	1.863
D^-	$\bar{d}c$	1.868
J/Ψ	$\bar{c}c$	3.097
B^-	$\bar{u}b$	5.26
B°	$\bar{d}b$	5.26
Υ	$\bar{b}b$	9.46

Important Baryons and Their Quark Structure

Baryon	Structure	Mass
p	uud	0.938 GeV
n	udd	0.940
Λ	uds	1.116
Σ^+	uus	1.189
Σ°	uds	1.192
Σ^-	dds	1.197
Ξ°	uss	1.315
Ξ^-	dss	1.321
Ω^-	sss	1.672
Λ_c^+	udc	2.273

Carriers of the Different Forms of Interaction Observed in Nature

Interaction	Mediator	Mass
Strong	Gluons	None
Electromagnetic	Photon	None
Weak	W boson	80 GeV*
	Z boson	90
Gravitational	Graviton	None

*These are predicted values.

Glossary

Alpha particle   The nucleus of the helium atom. It consists of two protons and two neutrons. Alpha particles are radiated by some radioactive substances (alpha radiation). Frequently abbreviated as α particle.

Antiparticle   A particle with the same mass and spin as the particle in question but with opposite electric charge, baryon number, lepton number, and so on. For every particle, there is a corresponding antiparticle. Certain purely neutral particles, such as photons and π^0 mesons, are their own antiparticles. The antineutrino is the antiparticle of the neutrino, the antiproton is the antiparticle of the proton, and so on. Antimatter consists of antiprotons, antineutrons, and antielectrons (normally called positrons).

Asymptotic freedom   The diminution of forces between quarks at short distances. A phenomenon of chromodynamics.

Baryons   A class of strongly interacting particles, including neutrons and the unstable hadrons known as hyperons. Baryon number is the total number of baryons present in a system minus the total number of antibaryons.

Beta decay   The decay of the neutron into a proton, an electron, and an antineutrino. This decay is a consequence of the weak

Glossary

interaction. Frequently abbreviated as β decay. The weak decay of a nucleus is also called β decay.

Boson   A concept for all particles that have integral spin. Examples of bosons are the π meson (spin 0), the photon (spin 1), and the W boson (spin 1).

Bubble chamber   An instrument for detecting particles. It consists of a vessel filled with a fluid heated nearly to the boiling point. When the pressure on the fluid is diminished suddenly, the fluid becomes overheated. At this moment electrically charged particles are shot through the bubble chamber. The fluid begins to boil along the paths of the particles, developing steam bubbles that can be photographed.

Charm   A quantum number that is equal to the number of charm quarks in a particle minus the number of anticharm quarks. The lightest particles with charm are the D mesons, which are approximately twice as heavy as the proton.

Charmonium   A system consisting of a charm quark and an anticharm quark.

Chromodynamics   A theory of the interaction between quarks and gluons. Physicists believe that quantum chromodynamics (QCD) is the correct theory of the strong interactions.

Conservation law   A law which states that the total value of some quantity does not change in any reaction.

Deuteron   The nucleus of deuterium. It consists of one proton and one neutron.

Electron   The lightest massive elementary particle. All chemical properties of atoms and molecules are determined by the electric interactions of electrons with each other and with the atomic nuclei.

Electron volt   A unit of energy equal to the energy acquired by one electron in passing through a voltage difference of 1 V. Equal to 1.602×10^{-19} watt·second. Often the units MeV ($= 10^6$ eV) and GeV ($= 10^9$ eV) are used.

Fermi constant A fundamental constant of nature which describes the strength of the weak interactions, denoted by G. Expressed in units of energy, $G = 294$ GeV^{-2}.

Fermion A generic term for all particles whose spin is 1/2.

Feynman diagrams Diagrams that show various contributions to the rate of an elementary-particle reaction. Introduced in the 1950s by Richard Feynman.

Fine structure constant The fundamental numerical constant of atomic physics and quantum electrodynamics, defined as the square of the charge of the electron divided by the product of Planck's constant and the speed of light. Denoted by α and equal to $1/137.036$.

Flavor An index that denotes the different types of quarks. One normally uses the symbols u, d, s, c, b, and t.

Gauge theories A class of field theories currently under intense study as possible theories of the weak, electromagnetic, and strong interactions. Such theories are invariant under a symmetry transformation, whose effect varies from point to point in space-time.

General relativity The theory of gravitation developed by Albert Einstein in the decade 1906–1916. As formulated by Einstein, the essential idea of general relativity is that gravitation is an effect of the curvature of the space-time continuum.

Glueball A neutral meson consisting solely of gluons.

Gluons Electrically neutral objects that mediate the interaction between quarks within the framework of chromodynamics. Gluons have spin 1.

Group A mathematical system of different (sometimes infinitely many) elements, with very definite rules. For example, the multiplication of two elements within the group is a well-defined operation. Group theory plays a very important role in physics. For example, it allows us to describe the symmetry of particles in a very simple manner.

Glossary

Hadron A particle that participates in the strong interaction. Hadrons are divided into baryons (such as the neutron and proton), which consist of three quarks each and obey the Pauli exclusion principle, and mesons, which do not obey this principle. Baryons have nonintegral spin $(1/2, 3/2, \ldots)$; mesons have integral spin $(0, 1, 2, \ldots)$.

Hydrogen The lightest and most abundant chemical element in the universe. The nucleus of ordinary hydrogen consists of a single proton. There are also two heavier isotopes, deuterium and tritium. Atoms of any sort of hydrogen consist of a hydrogen nucleus and a single electron.

Hyperons Baryons whose strangeness is nonzero, that is, baryons that contain at least one strange quark. The lightest hyperon is the Λ particle, with strangeness -1 and quark composition uds.

Infrared slavery A term which describes the confinement of quarks and gluons inside hadrons.

Jet A system of particles produced during particle reactions at high energies. The jets are interpreted as fragments of elementary objects such as quarks and gluons.

Leptons A class of particles that do not participate in the strong interactions, including the electron, muon, and neutrino. Lepton number is the total number of leptons present in a system minus the total number of antileptons. Leptons have spin $1/2$.

Maxwell's equations A group of equations that describe the dynamics of electromagnetic fields, derived by James Clerk Maxwell in the nineteenth century.

Mesons A class of strongly interacting particles, including the π mesons (also called pions), K mesons, ρ mesons, and so on, with zero baryon number.

Muon An unstable elementary particle of negative charge, similar to the electron but 207 times heavier; denoted by μ.

Neutrino A massless or nearly massless electrically neutral particle, denoted by ν. It participates only in weak and gravitational interactions and comes in at least three varieties, known as the electron-neutrino (ν_e), the muon-neutrino (ν_μ), and the tau-neutrino (ν_τ).

Neutron   The uncharged particle found along with protons in ordinary atomic nuclei; denoted by n.

Nucleon   Generic term for protons and neutrons.

Parity   A number that describes the symmetry of a system during reflection. If the system is symmetric, its parity is defined as $P = +1$; if antisymmetric $P = -1$. Parity is a constant in strong and electrodynamic interactions and does not change with time. Parity is violated in the weak interaction; that is, an object of positive parity can change with time into an object of negative parity, and vice versa.

Pauli exclusion principle   The principle that no two particles of the same type and with nonintegral spin can occupy precisely the same quantum state. Obeyed by baryons and leptons, but not by photons or mesons.

Photon   In the quantum theory of radiation, the particle associated with a light wave or, more generally, an electromagnetic wave; denoted by γ.

π Meson   The hadron of the lowest mass. Comes in three varieties, a positively charged particle (π^+), its negatively charged antiparticle (π^-), and a slightly lighter neutral particle (π^0). Sometimes called pion, the scientific notation is π meson.

Planck energy: The energy given by Newton's and Planck's constants: $Mp = (\hbar c / G_n)^{1/2}$. Its numerical value is $1.1 . 10^{19}$ GeV, which corresponds to 2.10^{-5} g in mass units.

Planck's constant   The fundamental constant of quantum mechanics; denoted by h. Planck's constant was first introduced in 1900, in Planck's theory of black-body radiation. It then appeared in Einstein's 1905 theory of photons: the energy of a photon is Planck's constant times the speed of light divided by the wavelength. Today the constant $\hbar$ is more frequently used, defined as Planck's constant divided by 2π. The numerical value of h is 6.6 X 10^{-34} watt · second^2.

Glossary

Positron The positively charged antiparticle of the electron; denoted by e^+.

Positronium The bound state consisting of an electron and a positron. Physicists differentiate between orthopositronium and parapositronium. In the ortho state, the spins point in the same direction; in the para state, they point in opposite directions.

Proton The positively charged particle found along with neutrons in ordinary atomic nuclei; denoted by p. The nucleus of hydrogen consists of one proton.

Quantum electrodynamics The quantum theory of electromagnetic phenomena (QED).

Quark The fundamental constituent of matter. All hadrons are supposed to be composed of quarks. Isolated quarks have never been observed, however, and there are theoretical reasons to suspect that, though in some sense real, quarks can never be observed as isolated particles.

Rest energy The energy of a particle at rest, which would be released if all the mass of the particle could be annihilated into radiation. Given by Einstein's formula $E = mc^2$.

Special relativity The view of space and time presented by Einstein in 1905. As in Newtonian mechanics, there is a set of mathematical transformations that relate the space-time coordinates used by different observers in such a way that the laws of Nature appear the same to all observers. In special relativity however, the space-time transformations have the essential property of leaving the speed of light unchanged, irrespective of the velocity of the observer. Any system containing particles with velocities near the speed of light is said to be relativistic and must be treated according to the rules of special relativity rather than the rules of Newtonian mechanics. In almost all elementary-particle reactions, it is necessary to use the rules of special relativity.

Speed of light The fundamental constant of special relativity, equal to 299 729 km/s; denoted by c. Any particles of zero mass, such as photons or neutrinos, travel at the speed of light. Massive particles approach the speed of light when their energy is very large relative to their rest energy.

Spin   A fundamental property of elementary particles that describes their state of rotation. According to the rules of quantum mechanics, spin can have only certain special values and is always either an integer or a half integer $(1/2, 3/2, 5/2, \ldots)$ multiplied by $\hbar$.

Strangeness   The quantum number relevant for the description of strange particles. Negative strangeness indicates how many s quarks are in a particular state.

Strong interaction   The strongest of the four general classes of elementary particle interactions. Responsible for the nuclear forces that hold protons and neutrons together in the atomic nucleus. The strong interaction affects only hadrons, not leptons or photons.

Vacuum polarization   The change of the physical properties of space in the neighborhood of a particle which interacts with other particles. For example, the space around an electron is filled with virtual positrons that influence the distribution of the electric charge of the electron.

W boson   The intermediary boson that mediates the weak interaction.

Wavelength   In any kind of wave, the distance between wave crests; denoted by λ. For electromagnetic waves, the wavelength may be defined as the distance between points where any component of the electric or magnetic field vector takes its maximum value.

Weak interaction   One of the four general classes of elementary-particle interactions. At ordinary energies, weak interactions are much weaker than electromagnetic or strong interactions, though very much stronger than gravitation. Weak interactions are responsible for the relatively slow decay of particles such as neutrons and muons. They are also responsible for all reactions involving neutrinos. It is now widely believed that the weak, the electromagnetic, and perhaps the strong interactions are different manifestations of a unified gauge field theory.

Z boson   A neutral heavy boson whose existence is predicted within the framework of the electroweak interaction. It mediates the neutral-current force.

References

The following brief list of books is by far incomplete. However, by consulting one or several of these books, readers may expand their knowledge of high-energy physics.

Aitchison, I., and Hey, I. *Gauge Theories in Particle Physics.* Bristol: A. Hilger Ltd., 1981.

Cheng, P. C., and O'Neill, G. *Elementary Particle Physics.* Reading, Mass. and London: Addison-Wesley, 1979.

Cheng, T. P., and Li, L. F. *Gauge Theory of Elementary Particles.* Oxford: Oxford University Press, 1983.

Close, F. *Quarks and Partons.* New York: Academic Press, 1978.

Frauenfelder, H., and Henley, E. *Subatomic Physics.* New York: Prentice Hall, 1974.

Lee, T. D. *Particle Physics and Introduction to Field Theory.* Chicago, London, New York: Harwood Academic Publishers, 1981.

Perkins, D. H. *Introduction to High Energy Physics.* Reading, Mass. and London: Addison-Wesley, 1972.

Taylor, J. C. *Gauge Theories of Weak Interactions.* Cambridge: Cambridge University Press, 1976.

The following articles dealing with particle physics that appeared in *Scientific American* and *Physics Today* may be of interest to readers.

Scientific American

Bloom, E. D., and Feldman, G. J. Quarkonium. May 1982.

Cline, D. B.; Mann, A. K.; and Rubbia, C. The Search for New Families of Elementary Particles. January 1976.

Cline, D. B.; Rubbia, C.; and van der Meer, S. The Search for Intermediate Vector Bosons. March 1982.

Drell, S. D. Electron-Positron Annihilation and the New Particles. June 1975.

Freedman, D. Z., and van Nieuwenhuizen, P. Supergravity and the Unification of the Laws of Physics. February 1978.

Georgi, H. A Unified Theory of Elementary Particles and Forces. April 1981.

Glashow, S. L. Quarks with Color and Flavor. October 1975.

't Hooft, G. Gauge Theories of the Forces between Elementary Particles. June 1980.

Jacob M., and Landshoff, P. The Inner Structure of the Proton. March 1980.

Johnson, K. A. The Bag Model of Quark Confinement. July 1979.

References

Lederman, L. M. The Upsilon Particle. October 1978.

Mann, A. K., and Rubbia, C. The Detection of Neutral Weak Currents. December 1974.

Nambu, Y. The Confinement of Quarks. November 1976.

Perl, M. L., and Kirk, W. T. Heavy Leptons. March 1978.

Schwitters, R. F. Fundamental Particles with Charm. October 1977.

Weinberg, S. Unified Theories of Elementary Particle Interaction. July 1974.

Wilczek, F. The Cosmic Asymmetry between Matter and Antimatter. December 1980.

Wilson, R. R. The Next Generation of Particle Accelerators. January 1980.

Physics Today

Cline, D., and Rubbia, C. Antiproton-Proton Colliders, Intermediate Bosons. August 1980.

Drell, S. D. When Is a Particle? June 1978.

Georgi, H., and Glashow, S. L. Unified Theory of Elementary Particle Forces. September 1980.

Heisenberg, W. The Nature of Elementary Particles. March 1976.

Turner, M. S., and Schramm, D. N. Cosmology and Elementary Particle Physics. September 1980.

Index

Index

and nucleons, 173, 175

Deuterons: antideuterons, 27; defined, 5, 276

Deutsches Electronen-Synchrotron (DESY), 13, 15, 61, 185, 190, 213, 216, 217, 266, 268

Dirac's equation and antimatter, 38

DORIS, 213

Einstein, Albert, 5, 7, 19, 27, 32, 259

Electric charge, *see* Electrodynamic forces, Electromagnetic forces, Electrons

Electrodynamic forces: and charges and colors of quarks, 135–37; and chromodynamics, 132–33; and chromomagnetic forces, 169–76; and gravity, 259; *see also* Electromagnetic forces, Electrons

Electromagnetic forces: and atomic decay, 47–49; and atomic particles, 47–64; and chromomagnetic forces, 169–76; as decay of positroniums, 27; Dirac's equation and antimatter, 38; early twentieth-century work on, 37–40; and electrically charged objects, 20–21; and energy levels of atoms, 22; and fields, as central concept in physics, 33; and nineteenth-century efforts, 31–36; and orthopositronium,

duration of, 38–40; and perturbation theory, 44–46; and quantum chromodynamics, 157–58; and strong interaction, 41–46; unified theory of, 31–40; and weak interaction, 224–28, 239–41

Electron accelerators, capacity of, 11

Electrons, 47, 171, 266; acceleration of, 11, 17, 77–81, 190; angular momentum of (spin), 22–25, 29–30, 170, 232; annihilation of, 147, 184, 185, 188, 189, 191, 208, 215; baryon number of, 57; defined, 276; electron volt, defined, 276; electric charge of, 19–20, 40; energy levels of, 22; and fine structure, 177–78; force between, at close distances and Coulomb's law, 147–49; magnetic moment calculation of, 173; mass of, 18–19, 146; movement of and classical physics, 131; movement of, and energy, 20; and neutron decay, 48; orbits of, 21; and photons, 140, 233; and positrons, 132, 139, 184, 185, 191, 210; and protons, compared, 19–20; quantum numbers of, 25; quantum states of, 22–23; and quark-antiquark pair, 135, 136; repulsion of, 35; in Rutherford model, 21; sliding off quark of, 179–82; spin of, 22–25, 29–30, 170, 232; and strong interaction, 42; and structure, 263; and u and d

Index

291

Index

Positrons, *(continued)*
of, 147, 184, 185, 188, 189,
191, 208, 215; and decay of
matter, 256; defined, 279;
and electrons, 132, 139, 184,
185, 191, 210; spin of, 29–
30; *see also* Positronium
Prediction in quantum theory,
6–7
Preons, 262
Protons, 5, 41, 47, 55, 56, 266;
acceleration of, 11; annihila-
tion/smashing of, 184, 193–
206, 220; and antiprotons,
27, 266; and decay, 6–7, 57–
58, 253, 254; defined, 280;
electric charge of, 19; and
electrons, compared, 19–20;
and exclusion principle, 25;
in hydrogen atoms, 20, 26;
life expectancy of, 250; and
mass, 77–78, 252; and neutrons, 65; and photons, 66;
and positronium, 26; proton-
antiproton collider, 206;
quark structure of, 80–82; in
Rutherford model, 21; stability of, 21, 57–60; and
strong interaction, 65–66;
velocity of and electromagnetic waves, 9; virtual, 36–
38; X-ray of, 76–87

QCD, *see* Quantum chromo-
dynamics theory
QED, *see* Quantum electrody-
namics theory
Quantum chromodynamics
(QCD) theory, 46, 138–55;

asymptotic freedom and
infrared slavery in, 165–66;
on chromoelectric confinement of color, 156–68; chromomagnetic forces in, 169–
75; and color analog of Coulomb's law, 159–60; and col-
or analog of QED, 141–45;
on color, magnetism, and
electricity, 157–58; and col-
or symmetry, 234–35; on
confinement, 154–68; de-
fined, 142; and electromagnetic forces, 157–58; and
fine structure of quarks,
178–82; and gluons, 208–20;
and interquark force, 202;
and unified (gauge) theory,
248–56; and vacuum, 145–
54; and weak forces, 221,
234–35; *see also* Quantum
electrodynamics theory
Quantum electrodynamics
(QED) theory, 5; and absolute certainty, 6–7; and Bohr
model, 131; color analog of,
141–45; and color analog of
Coulomb's law, 159–60; defined, 280; and electrons,
21–23; exclusion principle
in, 25; on natural events, 5–
6; on particle location and
velocity determination, con-
current, 7–8; on photons,
35–37; prediction in, 6–7;
professional acceptance of,
7; spin in, 23–25; and
strangeness, 90; and strong
force, 43–44; *see also* Quantum chromodynamics theory
Quarks, 79, 246, 273–74; and
antiquarks, 98, 135, 136,

Index